T0184270

Digital Social Networks and Travel Behaviour in Urban Environments

This book brings together conceptual and empirical insights to explore the interconnections between social networks based on Information and Communication Technologies (ICT) and travel behaviour in urban environments.

Over the past decade, rapid development of ICT has led to extensive social impacts and influence on travel and mobility patterns within urban spaces. A new field of research of digital social networks and travel behaviour is now emerging. This book presents state-of-the-art knowledge, cutting-edge research and integrated analysis methods from the fields of social networks, travel behaviour and urban analysis. It explores the challenges related to the question of how we can synchronize among social networks activities, transport means, intelligent communication/information technologies and the urban form.

This innovative book encourages multidisciplinary insights and fusion among three disciplines of social networks, travel behaviour and urban analysis. It offers new horizons for research and will be of interest to students and scholars studying mobilities, transport studies, urban geography, urban planning, the built environment and urban policy.

Pnina O. Plaut is a professor of transport and urban planning and policy at the Faculty of Architecture and Town Planning, Technion–IIT. Her research interests focus on smart cities and new mobility services and the nexus among ICT social networks, travel behaviour and urban environments. She is Initiator and Chair of the EU Horizon 2020 COST Action TU1305 titled: Social Networks and Travel Behaviour.

Dalit Shach-Pinsly is an architect and urban designer who is currently a senior lecturer at the Technion–IIT. Her research deals with measuring and evaluating diverse qualitative aspects of the urban environment, such as the Security Rating Index (SRI), presented in her article in *LAND* journal. She is Co-Initiator of the EU Horizon 2020 COST Action TU1305 titled: Social Networks and Travel Behaviour.

Transport and Society
Series Editor: John D Nelson

This series focuses on the impact of transport planning policy and implementation on the wider society and on the participation of the users. It discusses issues such as: gender and public transport, travel for the elderly and disabled, transport boycotts and the civil rights movement etc. Interdisciplinary in scope, linking transport studies with sociology, social welfare, cultural studies and psychology.

Published:

Transport Policy
Learning Lessons from History
Edited by Colin Divall, Julian Hine and Colin Pooley

The Mobilities Paradigm
Discourses and Ideologies
Edited by Marcel Endres, Katharina Manderscheid and Christophe Mincke

Ports as Capitalist Spaces
A Critical Analysis of Devolution and Development
Gordon Wilmsmeier and Jason Monios

Non-motorized transport integration into urban transport planning in Africa
Edited by Winnie Mitullah, Marianne Vanderschuren and Meleckidzedeck Khayesi

Forthcoming:

Visions, Concepts and Experiences of Travel Demand Management
Edited by Gerd Sammer, Michael Bell and Wafaa Saleh

Digital Social Networks and Travel Behaviour in Urban Environments
Edited by Pnina O. Plaut and Dalit Shach-Pinsly

Digital Social Networks and Travel Behaviour in Urban Environments

Edited by Pnina O. Plaut and Dalit Shach-Pinsly

Routledge
Taylor & Francis Group

LONDON AND NEW YORK

First published 2020
by Routledge
2 Park Square, Milton Park, Abingdon, Oxon OX14 4RN

and by Routledge
52 Vanderbilt Avenue, New York, NY 10017

Routledge is an imprint of the Taylor & Francis Group, an informa business

First issued in paperback 2021

British Library Cataloguing-in-Publication Data
A catalogue record for this book is available from the British Library

Library of Congress Cataloging-in-Publication Data
Names: Plaut, Pnina, editor. | Shach-Pinsly, Dalit, editor.
Title: Digital social networks and travel behaviour in urban environments / Pnina O. Plaut and Dalit Shach-Pinsly.
Description: Milton Park, Abingdon, Oxon ; New York, NY : Routledge, 2020. | Series: Transport and society | Includes bibliographical references and index.
Identifiers: LCCN 2019041438 (print) | LCCN 2019041439 (ebook) | ISBN 9781138594630 (hardback) | ISBN 9780429488719 (ebook)
Subjects: LCSH: Urban transportation. | Social networks. | City planning.
Classification: LCC HE305 .D55 2020 (print) | LCC HE305 (ebook) | DDC 307.2—dc23
LC record available at https://lccn.loc.gov/2019041438
LC ebook record available at https://lccn.loc.gov/2019041439

ISBN: 978-1-138-59463-0 (hbk)
ISBN: 978-1-03-208752-8 (pbk)
ISBN: 978-0-429-48871-9 (ebk)

Typeset in Times New Roman
by Apex CoVantage, LLC

We dedicate this book in tribute to the memory of our dear colleague Prof. Rein Ahas (1966–2018), a member and group leader in EU COST Action TU1305.

Contents

Tables

Figures

Contributors

Arnaud Adam is a PhD student in geography who has been hired by the BruNet project, where he studies the spatial organization of the Brussels area by means of community detection and new information and communication technology (ICT) data. He is also conducting sensitivity analyses of the community detection methods.

Giannis Adamos, PhD, graduated from the School of Engineering of University of Thessaly, Department of Civil Engineering, in 2004. In 2006, he acquired his MSc diploma in Applied Engineering and Systems Simulation and, in 2016, his PhD in Transportation Engineering from the Department of Civil Engineering of University of Thessaly. Since 2006, he is a scientific researcher at the Traffic, Transportation and Logistics Laboratory of University of Thessaly. From 2010 to 2015, he worked as a research associate at the Hellenic Institute of Transport of the Centre for Research and Technology Hellas in Thessaloniki, Greece, and from 2008 to 2011, he was a lecturer at the Technological Educational Institute of Larisa, Greece.

Rein Ahas, PhD (1966–2018), was Professor of Human Geography at the University of Tartu and founder of Mobility Lab, University of Tartu, Estonia. His studies mainly focused on the spatial mobility, urban studies, ICT and tourism and environmental change. His special field of interest was developing new methods for mobile phone–based research, concentrating to call detail records and smartphone-based active tracking. He was a member of the Eurostat Big Data working party and one of the initiators of the international conference series "Mobile Tartu". He was Research Professor of the Estonian Academy of Sciences from 2013 to 2015, and he was on the editorial board of the *Journal of Location Based Service* (Taylor & Francis), *Big Data and Society* (SAGE) and *Travel Behavior and Society* (Elsevier).

María del Mar Alonso-Almeida is Lecturer of Business Administration in Autonomous, Faculty of Economics and Business Administration, University of Madrid in Spain. She is interested in exploring the travel behaviour and social networks from a tourism viewpoint. She is interested in the study of transformations in destinations and organizations due to social media, as well as in the analysis of economic impact and the adaptability of new economic phenomenon.

Constantinos Antoniou is Full Professor and holds the chair of Transportation Systems Engineering at the Technical University of Munich, Germany. He has a diploma in Civil Engineering from the National Technical University of Athens (1995), an MS in Transportation (1997) and a PhD in Transportation Systems (2004), both from the Massachusetts Institute of Technology. His research focuses on modelling and simulation of transportation systems, intelligent transport systems, calibration and optimization applications, road safety and sustainable transport systems, and in his 20-plus years of experience, he has held key positions in a number of research projects in Europe, the United States and Asia, while he has also participated in a number of consulting projects.

Antònia Casellas is Professor and Researcher in the Geography Department at Universitat Autònoma of Barcelona (UAB) and Coordinator of the Master's Degree in Interdisciplinary Studies in Environmental, Economic and Social Sustainability at Institut de Ciència I Tecnologia Ambientals–UAB. Her research focuses on the interaction among economic viability, governance and urban morphology. She holds a PhD in Urban Planning and Policy Development, a Master in City and Regional Planning (MCRP) from Rutgers University (USA) and two bachelor degrees, in Philosophy from Universitat de Barcelona and in Communication Science from the UAB.

Emmanouil (Manos) Chaniotakis is Research Associate and PhD candidate at the Technical University of Munich. Until September 2016, he was a Research Associate at the Hellenic Institute of Transport of the Centre for Research and Technology Hellas. He holds a diploma in Rural and Surveying Engineering (2011) from the Aristotle University of Thessaloniki and an MSc in Transportation Infrastructure and Logistics (2014) from the Delft University of Technology. His research focuses on modelling and simulation of transportation systems, demand modelling and big data analysis in transportation.

Mario Cools holds a master's degree in applied economics, with a major quantitative business economics and a minor operations research (University of Antwerp, 2004) and a master's degree in applied statistics (Hasselt University, 2005). He is appointed as full-time associate professor at the Faculty of Applied Sciences of the University of Liège, where he is charge of research domain "diagnosis and analysis of transport and its externalities" of the research unit LEMA. LEMA is a research group attached to the Urban & Environmental Engineering research unit at the University of Liège, and develops its researches in the domains of (1) architectural and urban modelling with the goal to improve the sustainability of the built environment, (2) urban planning and urban forms and (3) diagnosis and analysis of transport and its externalities.

João de Abreu e Silva is Associate Professor at Instituto Superior Técnico, Universidade de Lisboa. His main research interests include travel behaviour and transport land-use interactions. João has participated in several COST Actions and in projects related with real-time data collection and data fusion. He has authored more than 40 articles in international peer-reviewed journals and has been involved in several international scientific networks, being currently member of the ADB10 and ADD30 Committees of the Transportation Research Board. He is Vice-Chair of the World Society for Transport and Land Use Research (WSTLUR) and a member of the Steering Committee for the Travel Survey Conference. He is editor of the *Journal of Transport and Land Use*.

Loukas Dimitriou is Assistant Professor in Department of Civil and Environmental Engineering, University of Cyprus (UCY) and Founder and Head of the LαB for Transport Engineering, UCY. His research interests focus in the application of advanced computational intelligence methods, concepts and techniques for understanding the complex phenomena involved in realistic transport systems and, further, in developing design and control strategies such as to optimize their performance. The methodological paradigms that he proposes utilize and/or combine elements from data science, behavioural analytics, complex systems modelling and advanced optimization and are applied to traditional fields of transport, such as demand modelling, travel behaviour and systems organization, optimization and control. He has more than 120 publications in peer-reviewed journals, proceedings of conferences and book chapters, while he is a regular reviewer in almost 50 international journals. Also, he an active member of international scientific organizations and committees.

Domokos Esztergar-Kiss is a research fellow at Budapest University of Technology and Economics. Since 2014 he is the international project coordinator of the Faculty of Transportation Engineering and Vehicle Engineering. His main fields of research are multimodal journey planning, personalized information services, activity-chain optimization, mobility as a service and sustainable mobility planning. He has organized the MT-ITS 2015 (Models and Technologies for Intelligent Transportation Systems), EWGT 2017 (Euro Working Group on Transportation meeting) and hEART 2019 (European Association for Research in Transportation) conferences. He is member of COST Actions TU1004 (Modelling Public Transport Passenger Flows in the Era of Intelligent Transport Systems), TU1305 (Social Networks and Travel Behaviour) and CA16222 (Wider Impacts and Scenario Evaluation of Autonomous and Connected Transport). He is involved in several Horizon 2020 and Interreg projects, such as MoveCit (Engaging employers from public bodies in establishing sustainable mobility), LinkingDanube (Transnational, multimodal traveller information and journey planning for environmentally friendly mobility in the Danube Region), MaaS4EU (Engaging employers from public bodies in establishing sustainable mobility and mobility planning) and electric travelling (platform to support the implementation of electromobility in smart cities based on information and communication technology [ICT] applications). He has prepared reviews for several international peer reviewed journals

(International Journal of Intelligent Transportation Systems Research, EURO Journal on Transportation and Logistics, Transportation Research Part A: Policy and Practice, Periodica Polytechnica Transportation Engineering).

Olivier Finance is a post-doctoral researcher in geography at UC Louvain (Center for Operations Research and Econometrics), hired in the BruNet project. He is developing an interactive visualization tool called atlas.brussels, to wrap up main results obtained during the project. He is moreover associate member at UMR8504 Géographie-cités (Paris).

Slaven Gasparovic is a geographer, and he works as Assistant Professor at the University of Zagreb, Faculty of Science, Department of Geography. He received his MA and PhD at the same university. He worked at several primary and secondary schools as a geography teacher. His research deals with transportation geography, especially with social aspect of transportation (i.e. transportation disadvantage, transportation based social exclusion, transportation poverty, transportation of vulnerable groups), as well as interrelation between ICT, social networks and travel behaviour. He has participated in several projects (e.g. CH4LLENGE, Sinergi, bilateral projects). He has been member of several COST Actions.

Maria Candelaria Gil-Fariña has been a full professor since 1997 in the Department of Applied Economics and Quantitative Methods – Universidad de la Laguna, Tenerife. A graduate in Economic Sciences and Management, with specialization in International and Regional Economics (1990), she has a PhD in Economic Sciences (1995). Her topics of research include Padé approximation in numerical analysis and multivariate time-series analysis; algebraic properties of Padé approximation and applications to economics; discrete and continuous time models applied to economics; the application of wavelets to economic (agricultural, energy, water) and financial (mortgages, gross domestic product) data; the estimation of water consumption in specific territories; sensitivity analysis for banana production according to the given incentives; and financial data modelling.

Bilin Han is a PhD candidate of the Urban Planning and Transportation Group of the Eindhoven University of Technology. Her research concerns modelling of work schedule arrangements and work participation in dual-earner households with children under social influence.

Matthew Hanchard is a Research Associate at the University of Glasgow working at the intersection of digital sociology, data science and urban studies. His current research with Beyond the Multiplex (AHRC: AH/P005780/1) focuses on developing a computational ontology from mixed-methods research to develop specialised film audiences. Matthew also acts as co-investigator for a Glasgow-Sydney University collaborative project called SmartPublics which focuses on the cultural, political, social and governance implications of 5G-enabled smart technologies in public spaces.

Bin Jiang is Professor in GeoInformatics and Computational Geography at University of Gävle, Sweden. He worked in the past with Hong Kong Polytechnic University, and University College London's Centre for Advanced Spatial Analysis. He is the primary developer of the software tool Axwoman for topological analysis of very large street networks. He invented the new classification scheme head/tail breaks for scaling analysis of big data. He is the founding chair of the International Cartographic Association Commission on Geospatial Analysis and Modeling. He used to be Associate Editor of international journal *Computer, Environment and Urban Systems* (2009–2014), and is currently Academic Editor of the open-access journal *PLOS ONE* and Associate Editor of *Cartographica*. His research interests centre on geospatial analysis and modelling of urban structure and dynamics, for example agent-based modelling, scaling hierarchy and topological analysis applied to street networks, cities and geospatial big data. Inspired by Christopher Alexander's work, he developed a mathematical model of beauty, which helps address why a design is beautiful, and how much beauty the design has.

Tom Erik Julsrud, PhD, is a sociologist and works as Senior Research Scientist at CICERO Center for International Climate Research and The Institute of Transport Economics (TOI) in Oslo. His current research interests include social practice theory, sustainable consumption, sociotechnical innovation theory, trust and collaborative consumption. He has recently published the book *Trust in network organizations*, at Fagbokforlaget (2018).

Sven Kesselring is a sociologist. He holds a professorship in Sustainable Mobilities at Nürtingen-Geislingen University, Germany (HfWU). He is the director of the Sustainable Mobilities master program at HfWU, the founder of the Cosmobilities Network, and its speaker since 2004. Cosmobilities is a global network of mobility researchers which promote social science based mobilities research. His main research areas are sustainable mobility policies, the transformations of automobilities, corporate mobility regimes and the social theory of the mobile risk society.

Jinhee Kim is Assistant Professor of the Department of Urban Planning and Engineering at Yonsei University, Republic of Korea. He was Assistant Professor of the Urban Planning and Transportation Group of the Eindhoven University of Technology, the Netherlands, until 2018. He earned a PhD in Urban Planning and Engineering, emphasizing on transportation planning from Yonsei University, August 2013. He has research interests in modelling behavioural process in activity-travel decisions, including the effects of social networks on decision-making process, and developing advanced discrete-choice models. He is a member of the International Association for Travel Behaviour Research and a life member of the Korean Society of Transportation.

Agnieszka Lukasiewicz (PhD candidate) is Researcher in Economic Division, the Road and Bridge Research Institute, Poland. She is an expert in stakeholders as well as managing large infrastructure projects analysis. Agnieszka's doctoral thesis is about financial dimensions of stakeholders' influence on large road infrastructure projects. She was also a project manager of national research and development project Collaboration of standard proceedings with stakeholders' relationships in the process of planning and realizing road and rail infrastructure projects. She has an experience in participation in various research projects related to transportation. She also has practice in cooperating with different kind of institutions, local authorities and informational documents preparation, especially in context of foreign direct investment (2000–2005 employed in Polish Information and Foreign Investment Agency). She also has experience in strategy building for enterprises and governmental institutions (e.g. Ukrainian Road Research Strategy project pro-investment strategy for Polish ICT sector, foreign investor relations based on electronic communication channels for Polish Information and Foreign Investment Agency and local authorities). She has taken part in many international projects, FP7, FP6, COST Actions and recently HORIZON 2020 (BENEFIT project).

Jasna Mariotti is Lecturer in Architecture at Queen's University Belfast. She studied architecture in Skopje and at Delft University of Technology, where she graduated with honours in 2007. In 2014 she completed a PhD on post-socialist cities and their urban transformations at the Faculty of Architecture, University of Ljubljana. Her current research focuses on the relationship between urban history, planning and architecture, urban governance and city retailing in the 20th and 21st centuries. Since 2010, Jasna Mariotti has worked on projects and urban interventions in different scales. Previously she was architect and urban designer in WEST 8 in Rotterdam, working on master plans and public space design.

Yufan Miao is a PhD researcher in the Future Cities Laboratory (FCL), Singapore–ETH Centre. He received his master's degree in Computational Science from Uppsala University and bachelor's in Geomatics from University of Gävle under the supervision of Prof. Bin Jiang. His current research interest is cognitive design computing as part of the Big-Data Informed Urban Design project in the research scenario of responsive cities in FCL. He aims to enrich and apply methods from cognitive computing to automate and optimise the urban design process for the facilities of urban designers.

Marija Mitrović Dankulov is Assistant Research Professor at the Scientific Computing Laboratory and deputy head of Innovation Centre at the Institute of Physics Belgrade, University of Belgrade. She has extensive knowledge and experience in theoretical and computational physics. Her main research interests are statistical physics of complex systems, with the emphasis on physics of socio-economic systems, and theory of complex networks. She completed her PhD in statistical physics at the Faculty of Physics, University of Belgrade in 2012. After her PhD studies, during which she was employed at the Department of Theoretical Physics, Institute Jožef Stefan, Slovenia, she undertook postdoctoral work at Department of Biomedical Engineering and Computational Science, Aalto University, Finland.

Gaëtan Montero, PhD student in geography (hired by the BruNet project). He particularly studies morphological issues and their link with social networks in the urban agglomeration of Brussels.

Itzhak Omer is Professor and Researcher in the Geography and Human Environment Department at Tel Aviv University, and also directs the Urban Space Analysis Laboratory, where he and his students work on the relationship between the urban environment and the spatial behaviour of individuals and developing of spatial analysis methods and simulation models, for studying movement and activities in the city. The areas of his academic interest include urban modelling, spatial behaviour and cognition, urban morphology, urban movement, urban systems and social geography of the city.

Pnina O. Plaut is Associate Professor of Transport and Urban Planning and Policy at the Faculty of Architecture and Town Planning, Technion, Israel Institute of Technology. The Vice Dean for Graduate Studies and Research, the former chair of the Graduate Program in Urban and Regional Planning and Director of the Center for City and Regional Studies at the Technion, Prof. Plaut holds a BSc in Civil Engineering, an MSc in Urban and Regional Planning from the Technion and a PhD in City and Regional Planning from the University of California, Berkeley. Prof. Plaut is the Initiator and Chair of the EU Horizon 2020 COST Action TU1305 titled: Social Networks and Travel Behaviour. Her fields of interests include transportation and land-use planning and policy, the impacts of infrastructure (transportation and telecommunications) on economic development and urban/regional structure and the environment. Her special research interests are smart cities and new mobility services and the nexus among ICT social networks, travel behaviour and urban environments.

Anniki Puura is a PhD student and Junior Research Fellow of Human Geography at the University of Tartu. Her research is focused on relationships between social networks and spatial mobility. In her studies, she mainly uses mobile phone–based data sets.

Soora Rasouli is Full Professor and Chair of the Urban Planning and Transportation Group, Eindhoven University of Technology. She received her master of Road and Transportation in 2005 at Tehran University and completed her PhD at the Eindhoven University of Technology. She specializes in developing travel behaviour models under uncertainty, regret-based models and social network analysis, in general, and (travel) behaviour, in particular.

Dalit Shach-Pinsly is an architect and urban designer who received her PhD at the Faculty of Architecture and Town Planning, Technion – IIT. She is currently a researcher and senior lecturer at the Technion – IIT. She was a Partner Investigator in the H2020-MG-9.2–2014 MIND-SETS – Mobility Innovations for a New Dawn in Sustainable Transport Systems. She is participating in the EU FP7 TUD COST Action TU1305 as the co-initiator and Management Committee member. She was a senior researcher of FP7-DESURBS-EU-Designing

Safer Urban Spaces. Formerly she joined the College of Built Environments at the University of Washington as a post-doctoral fellow sponsored by the competitive Marie Curie EU IOF Fellowship (2008–2011). Her fields of interests include built environment; measuring and evaluating diverse qualitative aspects of the urban environment, such as security and safety, visibility and privacy; master plan, analysis, methods and tools; and urban regeneration.

Fariya Sharmeen is a scholar of urban planning with specializations in built environment, travel behaviour and social networks. She is Assistant Professor of Sustainable Transportation at Radboud University, Nijmegen. Her research interests consist social contexts of urban and transportation planning and urban regeneration. Her PhD research was focused on the relationship between social network dynamics and travel behaviour. She is the recipient of Royal Geographic Society Postgraduate Award in 2013. Previously she worked as Lecturer at Bangladesh University of Engineering and Technology, Bangladesh, and as researcher at Cities Institute, UK, and Utrecht University, the Netherlands.

Siiri Silm, PhD, is Research Scientist of Human Geography at the University of Tartu and acting Head of Mobility Lab, University of Tartu. Her fields of research include human spatial-temporal behaviour, analyses of urban space, social networks and segregation (ethnic, age-related). She has participated since 2004 in developing mobile phone–based methodology and conducting research.

Francois Sprumont is a PhD student at the MobiLab Transport Research Group of the University of Luxembourg. After a bachelor in geography (Namur, Belgium), he obtained a master's degree in Spatial Planning from the University of Luxembourg. At that period (2009), he was among the 10 best students of the university and, as an award, participated to a one-month summer school in Jinan, China. More recently, he obtained his own grant from the FNR (Fonds national de la recherche) for the STABLE (Sustainable Transport behaviour considering Activity chains of BelvaL commutErs) research proposal.

Silvana Stefani is Full Professor of Mathematics Applied to Economics and Finance, Department of Statistics and Quantitative Methods, Università Milano – Bicocca, Italy. Her main research activities include discrete mathematics applied to economics and finance: financial graphs; spectral properties of matrices and graphs, networks and centrality; energy and environmental markets: modelling optimal production and hedging strategies in commodity and (conventional and renewable) energy markets, including mean reverting strategies in commodity markets; environmental markets and European Union Allowances; forward and futures premia; and ranking and journal classification, applying fuzzy statistical techniques to bibliometry.

Isabelle Thomas is Research Director at FRS-FNRS and full professor at the Université catholique de Louvain (Belgium). She is the 2016–2019 research director of the Centre of Operation Research and Econometrics (CORE). Her research fields are anchored in quantitative geography (economic/transport/ urban geography) often in a multidisciplinary context. She has (co-)authored numbers of publications in books and international scientific journals. She is and was involved in several national and international research projects and scientific commissions; she is member of the editorial board and referee for several journals in geography and spatial economics.

Harry Timmermans holds a PhD degree at the Catholic University of Nijmegen. Since 1976 he has been affiliated with the Department of the Built Environment of the Eindhoven University of Technology, the Netherlands. In 1985 he was appointed chaired professor of Urban Planning. His main research interests concern the study of human judgement and choice processes. In 2018, he received an honorary doctorate at Hasselt University, Belgium. Currently, he is also associated with the Department of Air Transportation Management, Nanjing University of Aeronautics and Astronautics.

Maria Tsami is Spatial Planning and Development Engineer (MSc) holding also an MSc in computer sciences. She is a scientific researcher and PhD candidate of the Traffic, Transportation and Logistics Laboratory at the Department of Civil Engineering of the University of Thessaly, and a scientific researcher of the Hellenic Institute of Transport. Her core research activity focuses on dynamic transit modelling, taking into account information and quality-of-service indicators.

Ann Verhetsel is Full Professor Economic Geography at the University of Antwerpen; she is Head of the Department of Transport and Regional Economics. She fulfils her appointment (research, education and services) in areas at the interface of geography, economics and planning. Ongoing research assignments under her leadership include projects and PhDs in the fields of mobility, smart specialization and new regional economic policy, networks within the logistics industry, the geographical dimension of venture capital, urban logistics and the location of retail.

Anne Vernez Moudon is Professor Emerita of Architecture, Landscape Architecture, and Urban Design and Planning; Adjunct Professor of Epidemiology and Civil and Environmental Engineering at the University of Washington, Seattle, where she also directs the Urban Form Lab. Dr. Moudon holds a BArch. (Honors) from the University of California, Berkeley, and a Doctor ès Science from the École Polytechnique Fédérale of Lausanne, Switzerland. Dr. Moudon was President of the International Seminar on Urban Morphology; Faculty Associate at the Lincoln Institute of Land Policy in Cambridge, MA; Fellow of the Urban Land Institute in Washington, D.C.; and a national advisor to the Robert Wood Johnson Foundation program on active living research. She is Professeur des Universités and Chercheur Associé, NEMESIS team of IPLESP (Pierre Louis Institute of Epidemiology and Public Health) in Paris. Dr. Moudon lectured at universities in Europe, Latin America and Asia. She published extensively in urban design, transportation and public health journals. Her books include *Built for Change: Neighborhood Architecture in San Francisco* (MIT Press 1986), *Public Streets for Public Use* (Columbia University Press 1991) and *Monitoring Land Supply with GIS* (Wiley & Sons 2000).

Bridgette Wessels is currently Professor in Social Inequality at the University of Glasgow (UK). Her research focuses on the innovation and use of digital services in social context. She has addressed the Internet and World Wide Web in everyday life, the public sphere, community life, telehealth, digital divides and social networks. Her books include *Communicative Civic-ness: Social Media and Political Culture* (Routledge 2018), *Open Data and Knowledge Society* (Amsterdam University Press 2017), *Understanding the Internet: A Socio-Cultural Perspective* (Palgrave 2010) and *Social Change: Process and Context* (Palgrave 2014).

Acknowledgement

We wish to thank the EU COST Association for funding the collaboration framework that enabled us to create this joint work. We also wish to thank Dr. Mikael Pero for his constant support in managing our COST Action TU1305. Special thanks to Ms. Smadar Amir for her wonderful work as Administrative Coordinator of the COST network and as Editorial Assistant of this book.

Introduction

Pnina O. Plaut and Dalit Shach-Pinsly

Over the past decade we have witnessed a rapid development of the information and communication technology (ICT), leading to extensive societal impacts. ICT enhanced the shift from social groups defined by location to individually based digital social networks. High-speed telecoms allow ad hoc personalized networks that affect general travel behaviour. The appearance of virtual social networks, such as Facebook and WhatsApp, and changes in working patterns (home/hub-based, shorter workdays) have resulted in the intertwining of leisure activities with other daily routines. All the preceding have an impact on travel and mobility patterns within the urban space.

A new field of research, ICT social networks and travel behaviour, is now emerging. It is focused on the challenges related to the question of how we can synchronize between social network activities, transport means, intelligent communication/information technologies and urban form. Studies on the relationships between social networks analysis and activity-travel modelling approaches began only between 2003 and 2006. Obviously, these studies are still initial, and the linkage between the spatial, social and travel characteristics of the trips still need further understanding. There is a lack of common ground, conceptual frameworks and research methodologies to cope with this new field of inquiry. At this stage, the relevant academic/research community is in need of consolidation of existing approaches, discussion in depth of theoretical and methodological issues, an exploration of practical applications, and the formulation of a research and policy agenda. This book aims to address these issues and the gap of knowledge in the field.

This book is a product of the work of the members in the European Union–funded COST Action TU1305 on social networks and travel behaviour. The COST Action initiated a new collaboration framework for 31 European research groups that explored the ways in which social activities, through ICT social networks, become mobilized in space and identified how social ties affect the integration of transport means into urban patterns. The group includes academic research teams with interdisciplinary expertise related to the subject, including sociologists, transportation researchers, information technology people, communications researchers, computer analysts, economist, physicist, geographers, architects, planners and urban analysts. The academic work presented in the book is not restricted to the EU and COST contributors. It is international and comprehensive. It includes references to the work of the leading scholars in the field.

The book presents state-of-the-art knowledge, cutting-edge research results and new perspectives. It integrates analysis methods from three disciplines, including social networks, travel behaviour and urban analysis. Each has its own theories, methodologies and tools. They built on previous research traditions and were developed parallel to one another. The societal impacts of rapid ICT developments on travel in urban areas have drawn attention to the lack of understanding about how these different research fields might be integrated. The book goes beyond the current literature by providing a platform for a broad scope of discussion regarding ICT social networks and travel patterns in urban space and, more important, by encouraging multidisciplinary fusion among these diverse disciplines. The book is composed of two parts. The first part (7 chapters) is focused on the theory and practice of ICT social networks and travel behaviour. The second part (5 chapters) is focused on the urban aspects of these relationships. The book ends with an epilogue.

The book addresses the scientific knowledge in the various disciplines aiming at drawing conclusions on (1) what can be learned from them in relation to the new interdisciplinary field of research which does not yet have profound theories and research methodologies and (2) how we can integrate relevant components and established practices into this field and create new models and practices. Several chapters (Chapters 1–4, 8–10) are focused on theories, models, methodologies and definitions and are based on areas of knowledge and disciplinary theories and practices. These are not related to a country base or from an EU perspective. Other chapters (Chapters 5–7, 11, 12) focus on specific European-based research and European data. However, the empirical results are discussed within the broad framework of theories, models and research practices, demonstrating the new role of social network and social media data. These raise methodological issues that are relevant to researchers and professionals worldwide.

The following describes the flow and the chapters:

Part I – theory and practice of ICT social networks and travel behaviour – includes Chapter 1, which focuses on social network theory, definitions and practice from various disciplines. The outcome is insights on what we have learned from other disciplines and whether and how they can contribute relevant knowledge to our field. Chapter 2 centres on social networks in the context of the mobilities theory, which provides a broad framework for research in the field. Chapter 3 presents a new perspective that comes from economic theories and introduces an opinion-dynamic model into our field. Chapter 4 focuses on social media and travel behaviour. Here, we explore the distinction between social media and social networks (which is somewhat blurred), make some observations regarding the two and concentrate on the relationship between social media usage and travel behaviour. Chapters 5 and 6 each present an empirical quantitative study while Chapter 7 presents an empirical qualitative study.

Part II of the book – the urban perspective of ICT social networks and travel behaviour – includes Chapter 8, which presents a typology of urban analysis models from three disciplines: geography, transportation and architecture. The analysis focuses on the potential to incorporate social networks into the models. Chapter 9, through a literature review of empirical studies, explores how social networks and their travel behaviour interact with the urban environment. Chapter 10 focuses on conceptualizing cities in light of the relations between social media and modes of travel. Chapter 11 presents an empirical quantitative study of urban analysis that is based on ICT data from Brussels (derived from smartphones) and discusses urban models in light of the study results. Chapter 12 presents the evolution of natural cities that emerges from an empirical quantitative study based on Twitter. The Epilogue concludes the book and is a synthesis of ideas, pointing to future directions of ICT enabled travel and discusses public concerns.

Both parts of the book are written by authors who come from different disciplines. The book is focused on the interdisciplinary nature of this new field of inquiry which is hard to capture. The integrative analysis sheds light on new insights and offers new horizons for research in the field regarding theories and methods. This is the strength of the book. We were fortunate to have the COST Action as a platform and succeeded in creating this kind of cooperation and joint analysis which would have been difficult to achieve elsewhere.

Part I

Information and communication technology social networks and travel behaviour – theory and practice

Part I

Information and
communication technology,
social networks and
(anti)social behaviour – theory
and practice

1 Social networks theory

Definitions and practice

Marija Mitrović Dankulov, María del Mar Alonso-Almeida, Fariya Sharmeen and Agnieszka Lukasiewicz

Introduction

Humans are social beings. Development and organization of societies and the behaviour of individuals are governed and shaped by social interactions (Wasserman & Faust, 1994). That is why it is not surprising that the structure and evolution of social networks are subject of various scientific disciplines, from sociology, economics and transportation studies to mathematics, computer science and physics. Sociologists introduced the term and the idea of social networks at the end of the 19th century to study the emergence of different social phenomena. Further development of this field during the first half of the twentieth century was led by scientists working in the field of psychology, anthropology and mathematics. During this period, scientists started with systematic recordings and analysis of social interactions in small groups, while mathematicians worked on developing a formalism to quantify the structure of social networks graph theory. The development of Information and Communication Technologies (ICT) at the end of the twentieth century and the availability of large data sets about human behaviour attracted the attention of physicists and computer scientists. These disciplines brought new concepts and powerful quantitative methods which further advanced our knowledge about the structure and dynamics of social networks (Sen & Chakrabarti, 2014).

This chapter provides a brief literature review of works on the topic of social networks in transportation studies, economics and physics. All these disciplines use a concept of social network and define it similarly. The differences between these disciplines are research problems and questions related to social networks, as well as their approaches. In the first part of the chapter, we provide a definition and classification scheme of social networks and describe some quantitative measurements of the network structure. In the second part of the chapter, we summarize the most relevant application of social network theory in transportation studies, economics and physics. We pay particular attention to the results, which indicate the connection between the structure of social networks and mobility patterns and travel behaviour of individuals.

Social networks theory

A social network is a theoretical concept used for the quantitative and qualitative description of social entities and relations between them. The social entities, actors, can be individuals, corporate or collective social units. A tie establishes a linkage between a pair of actors, and it can express a relation between two social entities like talking, kinship, friendship or business relations. In mathematical terms, a social network is a set of nodes (vertices), representing actors, and edges (links), representing social ties. Social ties can represent a direct relation, as in friendship or sexual partners' networks, or actors can interact indirectly, through artefacts, for example, a network of bloggers (Sen & Chakrabarti, 2014).

Social network analysis is used for exploring and quantifying patterns of relationships that arise among interacting social bodies, mostly individuals. An explicit assumption of such an approach is that indirect relationships in social groups matter. A particular backbone of social network analysis is that it provides standardized mathematical methods for calculating measures of sociality across levels of social organization, ranging from the population and group levels to the individual level (Freeman, 1984; McCowan et al., 2011). The concept of graphically representing social relationships is not new (Foster, Rapoport, & Orwant, 1963). Nevertheless, recent developments and widespread accessibility of network software have enabled easier visualization and exploration of complex social structures.

Social networks can be roughly divided into three classes based on the type of their ties: single-layer, temporal and multiplex networks (Boccaletti, Latora, Moreno, Chavez, & Hwang, 2006; Holme & Saramäki, 2012; Boccaletti et al., 2014). Single-layer networks are used for representation social systems whose actors interact through only one type of interaction, for example network of co-workers in one company. Depending on whether there are one or two types of actors, these networks can be monopartite or bipartite. Weighted networks are used to represent the systems where interactions can be of different strength. Finally, the interactions can be symmetric (undirected networks), as in the network of co-workers where the relationship is mutual or asymmetric (directed networks), where relationships are not reciprocal, such as student–mentor social network. Temporal networks are used for the representation of networks where ties and nodes are active at certain points in time (Holme & Saramäki, 2012), for example in mobile phone communication networks where phone call has a limited duration. Multiplex or multilayer networks are composed of a multiplicity of overlapping single-layer networks that capture different types of social connection, for instance actors can use different means of communication (phone calls, short message services, or online media) where each layer of a communication network has its own properties and dynamics. The size of this chapter does not allow us to cover all three types of network representations. A detailed description of methods and tools for the quantitative description of these networks can be found in review articles (Boccaletti et al., 2006; Holme & Saramäki, 2012; Boccaletti et al., 2014). Here we present only several quantitative measures used for the description of the topological structure of single-layer binary undirected networks: degree distribution, clustering coefficient and its dependence on node degree, of assortativity index, the dependence of average first neighbour degree on node degree and shortest path. It was shown that these properties are essential for the description of the topological structure of most real complex networks, including social networks (Orsini et al., 2015).

A quantitative description of social and complex networks requires the right set of tools. Graph theory is a natural framework for the mathematical representation of social and complex networks (Boccaletti et al., 2006). A network or a graph consists of two sets: a set of nodes (vertices) and a set of links (edges) that connects those nodes. Two connected nodes are said to be *adjacent* or *neighbouring*. The *node degree* is the number of its first neighbours.

One of the essential topological features of a network is degree distribution $P(q)$ defined as the probability that randomly chosen node has a degree q. The degree distribution is used for quantifying network heterogeneity at the local level and can be estimated as the fraction of nodes in the network having a degree q. The degree distribution is sufficient for a complete description of the structure of uncorrelated complex networks (Boccaletti et al., 2006; Orsini et al., 2015). However, most of the real, complex networks are correlated in the sense nodes with certain values of degree are more likely to be linked to each other. The degree correlations are characterized by conditional probability $P(q|q')$ which equals a probability that there is a link between nodes with degrees q and q'. The direct evaluation of conditional probability from the data is often not possible. Degree–degree correlations in a network can be estimated using average-nearest-neighbours degree and its dependence-on-node degree. For uncorrelated networks, the average-nearest-neighbours degree is independent of node degree. Correlated networks can be divided into two classes: assortative networks for which the average degree of nearest neighbours grows with q, and is disassortative where the opposite behaviour is observed.

Clustering, or transitivity, is another topological property of the networks which is particularly important for social networks. The clustering coefficient of a node is the probability that two randomly chosen neighbours of a node are also neighbours. It is estimated as a fraction of existing links out of all possible links between neighbours of a node. By averaging clustering coefficients over all nodes, one obtains the network clustering coefficient. Node and network clustering coefficients take the values between 0 and 1. Networks with a high value of clustering coefficient are considered to be clustered.

The shortest path has been one of the most important properties for characterization of network structure. A path between two nodes is an alternating sequence of nodes and edges, in which no node is visited more than once. The path of the minimal length is known as the shortest path. The shortest path of the largest length in the network is known as network diameter. The average shortest path of a network is defined as the mean of geodesic lengths over all pairs of nodes in the network. Most of the real, complex networks have relatively small average shortest-path length compared to their size, which is why they are often called small-world networks. A recent study (Orsini et al., 2015) has shown that small-world property can be explained using degree distribution, degree–degree correlation and dependence of clustering coefficient the on node degree.

Real complex networks are characterized by inhomogeneities on the mesoscopic level, also known as communities. The notion of community and the term itself have been proposed in the social sciences (Wasserman & Faust, 1994). In single-layer networks community is defined as a group of nodes more densely connected than with the rest of the network. The detection and quantitative description of the community structure of complex networks have attracted much attention in the past two decades (Fortunato, 2010). The networks can be characterized by different community structure. The community structure can be simple, with distinct communities. However, typical real-world networks have overlapping communities or communities that are hierarchically embedded. A detailed description of different algorithms and methods for finding different types of communities in static single-layered complex networks can be found in the paper by Fortunato (2010).

Social networks in various disciplines

Social networks in transportation studies

In the field of transportation, seminal empirical research on social networks has been documented. Among them are the studies by Wellman, Carrasco and colleagues (Carrasco, Hogan, Wellman, & Miller, 2008; Carrasco, Miller, & Wellman, 2008; Carrasco & Miller, 2005); Axhausen, Frei and Kowald (Axhausen, 2008; Frei & Axhausen, 2008; Kowald & Axhausen, 2012); and Timmermans, Van den Berg, Arentze, Sharmeen and colleagues (van den Berg, Arentze, & Timmermans, 2008, 2009, 2010, 2011, 2012; Sharmeen, Arentze, & Timmermans, 2013, 2014a, 2014b, 2015a, 2015b, 2016, 2017), based in Toronto, Zurich and Eindhoven, respectively. Most of those studies collected primary survey data, asking respondents to report a section of their social networks. Those studies came together into a comparative analysis of personal social network features in different spatial settings (Kowald et al., 2013) and were also combined in a recent book providing a much-needed overview of the relevant studies (Kowald & Axhausen, 2015).

Among the first attempts to understand social networks in transportation studies, is the connected lives study where name generators were employed to collect social network data (Carrasco & Miller, 2005; Carrasco et al., 2008a, 2008b). They focused on social activity generation, explicitly incorporating social network characteristics of each network member (alter) as well as the characteristics of the overall social structure. For a better understanding of the spatial distribution of social activity, they incorporated activity space anchor points, based in the home, institution and public spaces. Simultaneously, they characterized those places based on recurrence – whether these are regular places or not. The role of ICT on social interaction was also investigated.

On the other hand, Axhausen (2008) argued that social network membership influences a person's mental map, and therefore, the geographical network should have an impact on travel behaviour. In this study, he discussed in detail the survey instruments and data requirements for social network studies. In a subsequent study, Frei and Axhausen (2008) elaborated how geographical distances in personal social networks influence travel behaviour. They found that face-to-face contact frequency decreases with increasing distance whereas ICT (email) frequency increases.

Van den Berg et al. (2008, 2009, 2010, 2011, 2012, 2013) went in detail to investigate the interactions among ICT, socio-demographics, land use and social interactions. They reported a series of analyses of their social interaction diary and social network data collected in Eindhoven in 2008. The study employed a social interaction diary, followed by a name generator survey, to collect social network and interaction data. They extensively examined the impact of ICT on social travel behaviour and reported that the results differ significantly from a previous study conducted by Molin, Arentze, and Timmermans (2008), who used data about social networks collected in the 1980s, also in the Dutch context, implying that the inter-relations of social network and travel demand have changed over last two decades. They further investigated social travel for the elderly (2011) and the effect of club memberships on social interaction (2012). They reported that the elderly were as mobile as the young population in terms of frequency of social trips. The only difference in travel-mode choice was reported. They also delineated reciprocity in social network size and club memberships and, as expected, involvement in clubs and voluntary organizations increased social trips.

In most of the studies (including those mentioned earlier), social interaction diaries were commonly used, some in a different manner than the other. For example Silvis, Niemeier, and D'Souza (2006) used a similar social interaction diary, where instead of collecting information about network members, they provided an option for the network member to volunteer for the survey. Starting with three seed respondents, they collected information about 24 individuals over three phases of survey design. They concluded that individuals did not mind making longer trips for socializing and visiting family. They studied primarily concentrated on trip generation influence of social network. They also mapped the geography of trip frequencies.

In a similar direction, built environment effects of social interactions were investigated using time geography approach. In a recent study, Farber, Neutens, Carrasco, and Rojas (2014) calculated a Social Interaction Potential (SIP) metric that estimated the potential for an individual to engage in social activities given a specific time and space window. The metric was evaluated using data from cities Ghent and Concepcion. It was developed to assess the relationships between spatial structure and the potential opportunities for face-to-face contacts. The study provided a framework to assess the sociability of urban environments.

Sociability has been investigated from the perspective of solo versus joint activity planning. Ettema and Kwan (2010) analysed the company of social activities among ethnic groups in the Netherlands. They tested many hypotheses, contextual to social and recreational travel, and found that individuals had multiple networks (such as family, friends, associational and professional) which potentially perform multiple roles.

Thus far, the interplay between social networks and travel behaviour has been explored in travel behaviour research, in a particular space and time context. These studies are focused on static concepts of social networks, whereas social networks are dynamic. To develop a comprehensive understanding of network influence on individual/household's travel behaviour, it is imperative to examine those in a dynamic setting. Given the recency of the inclusion of social network components in travel behaviour studies, it is understandable the forthcoming arena of exploration.

The major setback of investigating social network evolution is perhaps data deficiency. Ideally, a panel study would be required, which is difficult with limited resources. Alternatively, social influences can be computed using stated preference surveys of expressed choices using non-linear utility functions (Kim, Rasouli, & Timmermans, 2014a, 2014b). Reviews of such efforts have also been detailed out in recent papers (Kim et al., 2018). The use of simulated or synthetic datasets has also shown potential to account for network dynamics. Dugundji and Gulyás (2008) aimed to incorporate social influence on transportation mode choice. They developed a multi-agent simulation model of household interactions, looking at how they decided on transportation mode alternatives by carefully distinguishing social and spatial network interdependencies. Páez, Scott, and Volz (2008) described a discrete choice model to account for social influence on decision making as an advancement over auto-correlation analysis. Social network simulation was developed based on the structure analysis tradition of sociology by developing an informal support network. In an earlier publication, Páez and Scott (2007) applied a similar methodology for decisions about telecommuting research. They pointed out certain limitations: first, the limited scope for empirical analysis due to limited mobility data with social network information and second, the dynamics of social networks were not taken into account. For example, although changing residential location meant new neighbours, schoolmates or gym mates, the social network was kept static in their model. Finally, they mentioned that they incorporated no influence from the "rest of the world". It is relatively difficult to incorporate that.

Han, Arentze and Timmermans (2011) presented a dynamic model that simulated habitual behaviour versus exploitation and exploration as a function of discrepancies between dynamic, context-dependent aspiration levels and expected utilities. Principles of social learning and knowledge transfer were used in modelling the impact of social networks, and related information exchange, adaptations of mutual choice sets and formation of common aspiration levels.

Dynamics concerning short-term decision making have been studied, as well as opening the scope for empirical analysis. Hackney and Axhausen (Axhausen & Hackney, 2006; Hackney, 2009) developed a multi-agent representation, incorporating dynamics of the social network, by addition and deletion links, based on feedback through activity-choice sets. They accounted for homophily and associations and assumed some maximum number of contacts per agent.

However, the structure of the social network and its characteristics have not yet been incorporated. To that end, Arentze and Timmermans (2008) developed a theoretical and modelling framework to capture the essence of social networks, social interactions and activity-travel behaviour. The core assumption was that the utility that a person derives from social interaction is a function of dynamic social and information needs, on one hand, and of similarity between the relevant characteristics of the persons involved, on the other. The model is consistent with the traditional social network theories (such as homophily and transitivity) developed in the social science literature. The process model has been tested using arbitrary agents. It led to the conclusion that an individual's social network had an equilibrium size dependent on several factors and changes over time. Although this study does formulate a theory and model of social network dynamics, no empirical data were collected to estimate the parameters of the model and test their theory. The same conclusion holds for Ronald, Dignum, Jonker, Arentze, and Timmermans (2012), which can best be seen of that line of work extension focusing on numerical simulation.

An elaboration of the framework is developed in Kowald and Axhausen (2012). Two Swiss data sets were used to simulate connected personal networks and encounters between actors. The model investigated leisure relationships and provided insights on the connectedness between actors and the factors affecting the leisure relationships between them. Illenberger, Flöteröd, Kowald, and Nagel (2009) conducted a similar simulation with a different approach. They tested network indicators, including edge-length distribution and network-degree distribution, in their model but did not account for properties such as homophily. Although those simulation frameworks were promising, most remain relatively simple. Thus, it remained a computational challenge to integrate large networks and complex social dynamics.

Hence, the dimensions and components of a social network may affect transportation choices in many ways, viz to perform joint activities, as a valuable source of information and social support and influencing daily as well as life choices (Sharmeen et al., 2015a). When social networks change, they potentially bring changes in any, or all, of the ways, eventually changing am individual's activity and travel patterns.

Seminal scholar Wellman and colleagues (Wellman, Wong, Tindall, & Nazer, 1997) reported a complete turnover in the network composition to those who got married during the study period. Furthermore, they outlined that continued telephone interaction and social support positively influenced tie maintenance. In a similar attempt of a qualitative study in three waves, Bidart and Lavenu (2005) found patterns of social network maintenance due to entry to the market, a start of a romantic relationship, childbirth and geographic mobility. Social interactions and relations to activity and travel schedule, however, remained largely unexplored. Furthermore, it is not readily evident how such research can be elaborated to fit the agenda of transportation research.

Only recently empirical studies of the influence of dynamics of social networks on activity and travel behaviour were reported (Chávez, Carrasco, Tudela, 2018;

Sharmeen, 2015; van den Berg, Weijs-Perrée, & Arentze, 2017). While, on one hand, time effects were investigated (Chávez et al., 2018), theories of life-cycle events to trigger changes were employed in the other (Sharmeen & Timmermans, 2014; Sharmeen et al., 2014a, 2014b, 2015a, 2015b; Sharmeen, Chávez, Carrasco, Arentze, & Tudela, 2016; Sharmeen & Sivakumar, 2017; Sharmeen, 2015; van den Berg et al., 2017). Through a series of studies, Sharmeen and colleagues documented the effects of life-cycle events on the interaction frequencies (Sharmeen et al., 2014b), mode choice (Sharmeen & Timmermans, 2014), time allocation (Sharmeen et al., 2013) and overall activity-travel need (Sharmeen et al., 2014a). Beyond those, they constructed a model to compute the population-wide evolution of social networks (Sharmeen et al., 2015b). This model was also replicated in other countries to predict the social network evolution (Sharmeen et al., 2016). The forecasting power and predictability of social contextual variables of individuals and positive correlation with different active travel dimensions were reported (Sharmeen & Sivakumar, 2017).

Empirical evidence from all these studies suggests that personal social networks do evolve with socio-demographic status and life-cycle events. Some ties got stronger and intertwined while others fade due to either a change in priorities or changes in an individual's or a household's time budget. Individuals alter and update their choices under the social influence or at specific points in life. Those points or life-cycle events act as triggers to deliberate choice decisions. Such decisions are better understood when the broader contextual environment of individuals is incorporated. Mainly, the social network composition and geography are overall found to be primarily associated with all aspects of activity-travel scheduling (and residential location choice). Therefore, social network attributes are found to offer a better understanding and are crucial for incorporation in large-scale travel demand forecasting models not only at any one point of time and space but also during the progression and consecutive changes over the short and long term. Together with the growing leisure travel duration and location choices and the explosion of social media, social networks have become increasingly crucial in travel behaviour research. Fuelled by the demand landscape, the field has gained momentum in recent years in transportation research summarized earlier. There is much to be done to realize the full potential of social influence in travel predictions.

Social networks explanation in the transportation field is through the character of social activities, their spatial distribution and frequency, and related travel behaviour. They are the driver behind the activity-travel decisions and a key source of explanation of activity-travel generation (Carrasco & Miller, 2009; Calastri, Hess, Daly, & Carrasco, 2017). Nevertheless, research on the effects of social networks on activity-travel patterns emerged for understanding social travel, in its own right, better and for improving the performance of comprehensive activity-based models of travel demand forecasting in which the prediction of social travel was a weak link. Over the last decade, numerous empirical studies have confirmed the contention that the intensity and nature of social activity-travel behaviour are significantly influenced by the properties of personal social networks (Kim et al., 2018). Thus, while the activity-based modelling has shifted from cross-sectional to multiple-horizon dynamic models (Sharmeen et al., 2015b; Rasouli & Timmermans, 2014), there is still space of further examination of the subject.

Social networks in economics

Social networks theory is widely applied, including economics and travel studies. There is the economic problem of scarcity in which individuals must face trade-offs while making choices. The available resources are insufficient to give satisfaction to all human wants. Choices made by persons determine the allocation of scarce resources. Social interactions and social networks are useful for examining how such decisions are made and what is the trigger off.

The social context where many economic interactions arise is taken into consideration as a significant factor and the driver of behaviours, decisions and outcomes. Social networks are used in obtaining jobs. They have an inevitable impact on decisions about buying products, education choice and so on. In the field of economic sociology, after Granovetter (1985) published a paper, the new economic sociology aroused. Thus, Granovetter has been connected to the idea of embeddedness. In such the economic relations between individuals or firms are embedded in actual social networks, as well as they do not continue in an abstract, idealized market. Granovetter (1977) stressed the importance of weak ties, more casual acquaintances outside the group of closest friends, for the job-seeking process. Montgomery (1991) digested the findings of various surveys coming up to the conclusion that about half of all currently employed workers found their job through information provided by friends or family members.

The late attentiveness of economists concerning social networks possibly comes from having pushed many economic models to their limits and found that social circumstances can help explain observed economic phenomena in ways that restricted economic models cannot. Individuals are assumed to form or maintain relationships that they find beneficial and avoid or remove themselves from relationships that are not beneficial. That is sometimes captured through equilibrium notions of network formation but is also modelled through various dynamics, as well as agent-based models where specific heuristics are specified that govern behaviour. That choice perspective traces the structure and the properties of networks back to the costs and benefits that they bestow upon their participants (Jackson, 2007).

There is a rising interest in social issues noticed in economics. The approach of new applied models can be connected with fairness criteria and existing inequalities among different groups, for example regarding employment or wages. In such conditions, the problem of market efficiency comes as the social context can change the ways the resources could be allocated. The attitude one can concern coming to the relatively new concept which is collaborative, sharing economy, in which social networks are successfully applied. The sharing economy, which is a fast-growing sector, attracts the attention of policy-makers, as sharing is related to democracy and seems to involve openness, inclusion and equality and is based on social networks and, at some point, could not exist without social media.

In the field of economics, models of network formation based on stochastic algorithms are used. Economists also model the formation of social networks as the outcome of optimization decisions. Social relations are connected with costs and benefits, and in the process of making a decision, the elements are weighted against each other. Those models make it possible to analyse the consequences of technological changes that alter the cost of communication (Mayer, 2009).

Social network analysis in economics has also been used to understand complex economics or management problems and make optimal decisions. Thus, there are some economics and management fields where social networks empirical analysis has been widely developed, such as the following:

- Operations where efficiency measures are used (Sadri, Ukkusuri, & Gladwin, 2017; Blas, Martin, & Gonzalez, 2018)
- The power of diversity in organizations regarding creativity, productivity or knowledge sharing among others (Park, Im, & Sung, 2017; Goyal, Rosenkranz, Weitzel, & Buskens, 2017; Chua, 2018)
- Organizations' collaboration (Shane & Cable, 2002; Chua, 2018; Boschet & Rambonilaza, 2018)
- Finance issues (Mollick, 2014; Huang & Knight, 2017; Polzin, Toxopeus, & Stam, 2018)
- Inversion issues (Garlaschelli, Battiston, Castri, Servedio, & Caldarelli, 2005; Wang & Wang, 2018)
- International trade flows (Barigozzi, Fagiolo, & Garlaschelli, 2010; Barigozzi, Fagiolo, & Mangioni, 2011; Lovrić, Da Re, Vidale, Pettenella, & Mavsar, 2018)

According to Boccaletti et al. (2006, p. 176), *the extensive and comparative analysis of networks from different fields has produced a series of unexpected and dramatic results*. Therefore, another theory to surplus the limitations of social network theory was needed. Thus, the authors, such as those mentioned earlier, explain the research on complex networks begun with the *effort of defining new concepts and measures to characterize the topology of real networks*.

Complex networks theory, according to some authors, provides a framework that can explain how changes in context, economics or human behaviour happen and which way the evolution could be. The theory explains that all networks share common properties. Therefore, applying complex network theory in economics makes possible identification and measure of social networks in this field and make decisions on the matter. Besides, this theory could shed light on how countries, firms or people interact and relate to themselves.

The mobility paradigm states that all places are linked into thin (or fat) networks of connections that stretch beyond each place (Sheller & Urry, 2006). In other words, nowhere can be an island (Gogia, 2006). Mobility, in a broad sense, also includes movements of income, information and images on local and global ways and one-to-one in different ways of communication, such as phones, smartphones and social media. Empirical analysis with this theory can use qualitative and quantitative methods (Hollstein, 2011).

For example, in the case of tourism, activities are not separated from the places visited. Indeed, the places travelled used to depend, at least in part, on activities can be practised within them. Authors mentioned earlier asserted that exists a complex relationship between the mean and the way of travel and the traveller. Thus, it is necessary to examine the topology of social networks and the patterns of weak ties that could generate small-world property (Urry, 2003, 2004).

Nevertheless, a specific theory which explains complex patterns forms and changes does not exist yet. For that reason, some authors suggest that economics involves the analysis of complex systems that are neither perfectly ordered nor anarchic (Capra, 2004; Sheller & Urry, 2006).

Social network theory can help to identify "critical networks" for example, main stakeholders in a specific place. Besides, it can provide action patterns and the role of every member in the network (Timur & Getz, 2008; Scott, Baggio, & Cooper, 2008). Moreover, complex network theory provides a robust foundation for identifying critical stakeholders and their relationships inside and offside of the network (Timur & Getz, 2008).

Boccaletti et al. (2006) define a complex system that comprises a large number of components usually non-linear and operated in a non-predictive way. In a tourism destination, it is possible to find a vast number of components, both tangible and intangible (Alonso-Almeida & Celemin-Pedroche, 2016). A complex adaptive system is continuously interacting with the environment and generating dynamic adjustments on the structure and behaviour. Therefore, a theory which studies complex systems such as complex network theory will be able to predict future conditions based on past trends (Andersen & Sornette, 2005).

Complex network theory has been built from observation of the real world network properties and structures (Miguéns & Mendes, 2008). In economics, this theory has been applied to ecological networks, financial relations and companies' collaboration mainly (Caldarelli, Battiston, Garlaschelli, & Catanzaro, 2004; Tibély, Onnela, Saramäki, Kaski, & Kertész, 2006).

Statistical physics of social networks

Different systems and macroscopic, collective, phenomena require different models for the description of their dynamics. However, there is one common feature present in all many-body systems, regardless of their nature: the underlying network of interactions between elements that constitute them. The empirical analysis of data collected for different biological and sociotechnical systems has shown that complex and heterogeneous connectivity patterns, typically found in those systems, are one of the key signatures of their self-organizing dynamics (Boccaletti et al., 2006; Holme & Saramäki, 2012; Boccaletti et al., 2014). In the past few decades, physicists have put much effort into understanding the mutual influence between the structure of the interaction network and system dynamics. They developed a set of quantitative methods and measures that allowed them to study this relationship in detail and better understand the emergence of various phenomena in social systems (Sen & Chakrabarti, 2014; Boccaletti et al., 2006; Holme & Saramäki, 2012; Boccaletti et al., 2014; Castellano, Fortunato, & Loreto, 2009).

Quantitative analysis of the structure of different social networks has shown that several universal features characterize all of them. Famous Milgram experiment has shown that an average number of connections between any two humans in the United States equals six, and that world is, in general, small (Travers & Milgram, 1967). A recent analysis of the shortest paths in Facebook has indicated that the world is even smaller than it was expected (Backstrom, Boldi, Rosa, Ugander, & Vigna, 2012). In a mathematical sense, small-world property means that the average shortest path in the network grows logarithmically with its size (Sen & Chakrabarti, 2014; Boccaletti et al., 2006). Many real-world networks are small world. Social networks are heterogeneous, on both the local and mesoscopic scales. On the local scale, that heterogeneity is manifested in long-tail degree distribution. Individuals differ in the number of friends and acquaintances they have: many of them have only a few friends, while still there are some that have a large number of friends and acquaintances. Societies and social groups are not homogeneous; that is they always consist of smaller groups of people who have more connections with each other than with the rest of the network. Social networks exhibit various types of community structure, from well-defined and separated communities to overlapping communities and communities which can be nested and thus form a hierarchical structure (Fortunato, 2010). Communities can grow, shrink and disappear (Palla, Barabási, & Vicsek, 2007). The size of the community has a crucial role for its dynamics and survival: smaller communities have longer survival time, and their membership is very stable over the long period, while the only way for a large community to survive is to change its members regularly.

Social networks exhibit positive degree–degree correlations. While friendship, collaborative and online social networks are assortative (Newman, 2002), bipartite networks, which represent social dynamics in different techno-social networks, are characterized with weak disassortative mixing (Mitrović & Tadić, 2012). A high value for the clustering coefficient is also a prominent feature of social networks (Boccaletti et al., 2006). The knowledge of how the basic properties emerge in real social networks thus is essential for understanding the evolution and dynamics of related social system and/or phenomena.

Theoretical models of complex networks have proved to be an invaluable tool for studying the evolution of social and complex systems. In parallel with the quantitative exploration of the structure of real social networks, physicists have worked on constructing theoretical models of evolving networks (Boccaletti et al., 2006; Holme & Saramäki, 2012; Boccaletti et al., 2014). Those models, although simplistic, can mimic the properties of real networks. These models enabled a detailed analysis of the emergence of different topological properties in networks and understanding of the primary mechanism that underlie social network evolution. Besides, models enable a comprehensive study on how and to which extent the network structure influences dynamical processes and the emergence of macroscopic phenomena. We provide a brief description of some of the most fundamental models.

The first and basic model in complex network theory is Erdős- Rényi (ER) random graph model (Erdős & Rényi, 1960). The degree distribution of ER graphs is Poasonian, and they belong to a class of uncorrelated graphs. The value of their clustering coefficient is very low and tends to zero as network size tends to infinity. The ER model is instrumental and irreplaceable as a null model in testing hypothesis related to complex networks. Watts–Strogatz (WS) model, or small-world, model was the first model that successfully reproduced a small diameter and large clustering coefficient of social networks (Watts & Strogatz, 1998). The properties of networks strongly depend on the value of the single parameter that controls the percentage of rewired links. Network without rewired links is a regular graph, while the graph obtained after rewiring of all links is equivalent to ER graph. Graphs with more than 0% and less than 100% of rewired links have a small-world property and a high value of clustering coefficients. However, WS networks do not have a broad degree distribution, one of the most prominent features of many real-world networks, but instead their degree distribution exhibits exponential decay. In order to reproduce the property, Barabasi and Alber (BA) proposed an evolving model of networks (Barabási & Albert, 1999). In the BA model, networks grow by following the preferential attachment rule. In the preferential attachment mechanism, the probability for an old node to be chosen as a target of a new link is proportional to the number of its previous connections. The obtained network has broad, power-law degree distribution, with the exponent equal to three in the limit of infinite size network. Although the average shortest path of these network grows logarithmically with the network size, the clustering coefficient tends towards zero when network size tends to infinity. Other models have been proposed, including different modifications of the BA model, in order to reproduce other properties of real-world social networks. Some of these models are based on the fact that many social networks are embedded in Euclidean space (Barthélemy, 2011), while others take into account different temporal factors, for instance, ageing (Hajra & Sen, 2005).

Network structure has a crucial role in the emergence of different collective states in social systems. The real-world and model-generated networks have been used for studying dynamical processes and their connection with network structure. Opinion dynamics, a process of collective opinion emerging in the social systems, in the convergence time and possibility to reach the consensus in the social group, depend on the topological properties of its underlying social network (Sen & Chakrabarti, 2014). Local heterogeneity of social networks plays a vital role in how and to which extent the disease spread through it. The quantitative methods of complex networks have proved very useful for identifying the individuals that play a crucial role in disease spreading and create the most effective immunization plans (Pastor-Satorras, Castellano, Van Mieghem, & Vespignani, 2015). The structure of social contacts influences long-distance travel of humans (Cho, Myers, & Leskovec, 2011) and can be used for predicting and modelling mobility patterns (Palchykov, Mitrović, Jo, Saramäki, & Pan, 2014). Social networks are not fixed. They evolve and are influenced by dynamics and activity in social systems. The exchange of emotions has been found to be an essential factor in life and death of online social communities (Mitrović & Tadić, 2012).

Conclusion

Social interactions shape and determine the evolution of human society. In that sense, social network theory, developed in parallel by different scientific disciplines, provided necessary tools for understanding the emergence of various social phenomena. This chapter provides an overview of the most important methods and results related to social networks in transport, economy and physics. Results from transportation studies show that social networks have a significant influence on every aspect of human transportation and travel behaviour. It has been shown that the length and duration of the trips of individual depend on the structure of its social contacts. However, the question of mutual dependence between the social network and travelling behaviour still stays an open question. Transportation and travel behaviour are inevitably significant and are one of the many activities that are influenced by social media and networks. Economic studies have shown that social contacts have a crucial role in the socio-economic development of any human society and that social networks are an essential component of both empirical and theoretical studies in the economy. The concept of complex networks, a broader class of studied systems that includes social networks, has originated from physics. A large number of empirical studies from physics has shown that social networks can be characterized with a relatively small set of properties: they are heterogeneous at the local and mesoscopic scales, assortative, cluster and have small-world property. Empirical studies, in combination with theoretical modelling, have shown that social networks determine the emergence of many collective phenomena in social systems, including travelling behaviour.

References

Alonso-Almeida, M. M., & Celemin-Pedroche, M. S. (2016). Competitiveness and sustainability of tourist destinations. *ESIC Market, 47*(2), 275–289.

Andersen, J. V., & Sornette, D. (2005). A mechanism for pockets of predictability in complex adaptive systems. *EPL (Europhysics Letters), 70*(5), 697.

Arentze, T., & Timmermans, H. (2008). Social networks, social interactions, and activity-travel behavior: A framework for microsimulation. *Environment and Planning B: Planning and Design, 35*(6), 1012–1027.

Axhausen, K. W. (2008). Social networks, mobility biographies, and travel: Survey challenges. *Environment and Planning B: Planning and Design, 35*(6), 981–996.

Axhausen, K. W., & Hackney, J. K. (2006). An agent model of social network and travel behavior interdependence. *Arbeitsbericht Verkehrs-und Raumplanung, 380*.

Backstrom, L., Boldi, P., Rosa, M., Ugander, J., & Vigna, S. (2012, June). Four degrees of separation. In *Proceedings of the 4th Annual ACM Web Science Conference*, Evanston, IL, (pp. 33–42). ACM.

Barabási, A. L., & Albert, R. (1999). Emergence of scaling in random networks. *Science*, *286*(5439), 509–512.

Barigozzi, M., Fagiolo, G., & Garlaschelli, D. (2010). Multinetwork of international trade: A commodity-specific analysis. *Physical Review E*, *81*(4), 46–104.

Barigozzi, M., Fagiolo, G., & Mangioni, G. (2011). Identifying the community structure of the international-trade multi-network. *Physica A: Statistical Mechanics and Its Applications*, *390*(11), 2051–2066.

Barthélemy, M. (2011). Spatial networks. *Physics Reports*, *499*(1–3), 1–101.

Bidart, C., & Lavenu, D. (2005). Evolutions of personal networks and life events. *Social Networks*, *27*(4), 359–376.

Blas, C. S., Martin, J. S., & Gonzalez, D. G. (2018). Combined social networks and data envelopment analysis for ranking. *European Journal of Operational Research*, *266*(3), 990–999.

Boccaletti, S., Bianconi, G., Criado, R., Del Genio, C. I., Gómez-Gardenes, J., Romance, M. . . . Zanin, M. (2014). The structure and dynamics of multilayer networks. *Physics Reports*, *544*(1), 1–122.

Boccaletti, S., Latora, V., Moreno, Y., Chavez, M., & Hwang, D. U. (2006). Complex networks: Structure and dynamics. *Physics Reports*, *424*(4–5), 175–308.

Boschet, C., & Rambonilaza, T. (2018). Collaborative environmental governance and transaction costs in partnerships: Evidence from a social network approach to water management in France. *Journal of Environmental Planning and Management*, *61*(1), 105–123.

Calastri, C., Hess, S., Daly, A., & Carrasco, J. A. (2017). Does the social context help with understanding and predicting the choice of activity type and duration? An application of the Multiple Discrete-Continuous Nested Extreme Value model to activity diary data. *Transportation Research Part A: Policy and Practice*, *104*, 1–20.

Caldarelli, G., Battiston, S., Garlaschelli, D., & Catanzaro, M. (2004). Emergence of complexity in financial networks. In *Complex networks* (pp. 399–423). Berlin; Heidelberg: Springer.

Capra, F. (2004). *The hidden connections: A science for sustainable living*. New York, NY: Anchor.

Carrasco, J. A., Hogan, B., Wellman, B., & Miller, E. J. (2008). Agency in social activity interactions: The role of social networks in time and space. *Tijdschrift voor economische en sociale geografie*, *99*(5), 562–583.

Carrasco, J. A., & Miller, E. J. (2005). Socializing with people and not places: Modelling social activities explicitly incorporation social networks. Paper presented at the Conference on Computers in Urban Planning and Urban Management, London.

Carrasco, J. A., & Miller, E. J. (2009). The social dimension in action: A multilevel, personal networks model of social activity frequency between individuals. *Transportation Research Part A: Policy and Practice*, *43*(1), 90–104.

Carrasco, J. A., Miller, E. J., & Wellman, B. (2008). How far and with whom do people socialize? Empirical evidence about distance between social network members. *Transportation Research Record*, *2076*(1), 114–122.

Castellano, C., Fortunato, S., & Loreto, V. (2009). Statistical physics of social dynamics. *Reviews of Modern Physics*, *81*(2), 591.

Chávez, Ó., Carrasco, J. A., & Tudela, A. (2018). Social activity-travel dynamics with core contacts: Evidence from a two-wave personal network data. *Transportation Letters*, *10*(6), 333–342.

Cho, E., Myers, S. A., & Leskovec, J. (2011, August). Friendship and mobility: User movement in location-based social networks. In *Proceedings of the 17th ACM SIGKDD International Conference on Knowledge Discovery and Data Mining*, San Diego, CA (pp. 1082–1090). ACM.

Chua, R. Y. (2018). Innovating at cultural crossroads: How multicultural social networks promote idea flow and creativity. *Journal of Management, 44*(3), 1119–1146.

Dugundji, E. R., & Gulyás, L. (2008). Sociodynamic discrete choice on networks in space: Impacts of agent heterogeneity on emergent outcomes. *Environment and Planning B: Planning and Design, 35*(6), 1028–1054.

Erdős, P., & Rényi, A. (1960). On the evolution of random graphs. *Publication of the Mathematical Institute of the Hungarian Academy of Sciences, 5*(1), 17–60.

Ettema, D., & Kwan, M. P. (2010, July). The influence of social ties on social and recreational activity participation of ethnic groups in the Netherlands. Paper presented at 12th WCTR, July 11–15, 2010, Lisbon, Portugal.

Farber, S., Neutens, T., Carrasco, J. A., & Rojas, C. (2014). Social interaction potential and the spatial distribution of face-to-face social interactions. *Environment and Planning B: Planning and Design, 41*(6), 960–976.

Fortunato, S. (2010). Community detection in graphs. *Physics Reports, 486*(3–5), 75–174.

Foster, C. C., Rapoport, A., & Orwant, C. J. (1963). A study of a large sociogram II: Elimination of free parameters. *Behavioral Science, 8*(1), 56–65.

Freeman, L. C. (1984). Turning a profit from mathematics: The case of social networks. *Journal of Mathematical Sociology, 10*(3–4), 343–360.

Frei, A., & Axhausen, K. W. (2008). Modelling the frequency of contacts in a shrunken world. *Arbeitsberichte Verkehrs-und Raumplanung, 532.*

Garlaschelli, D., Battiston, S., Castri, M., Servedio, V. D., & Caldarelli, G. (2005). The scale-free topology of market investments. *Physica A: Statistical Mechanics and Its Applications, 350*(2–4), 491–499.

Gogia, N. (2006). Unpacking corporeal mobilities: The global voyages of labour and leisure. *Environment and Planning A, 38*(2), 359–375.

Goyal, S., Rosenkranz, S., Weitzel, U., & Buskens, V. (2017). Information acquisition and exchange in social networks. *The Economic Journal, 127*(606), 2302–2331.

Granovetter, M. S. (1977). The strength of weak ties. *Social Networks*, 347–367.

Granovetter, M. S. (1985). Economic action and social structure: The problem of embeddedness. *American Journal of Sociology, 91*(3), 481–510.

Hackney, J. K. (2009). *Integration of social networks in a large-scale travel behavior microsimulation*. Dissertation, Eidgenössische Technische Hochschule ETH Zürich, Nr. 18723.

Hajra, K. B., & Sen, P. (2005). Aging in citation networks. *Physica A: Statistical Mechanics and Its Applications, 346*(1–2), 44–48.

Han, Q., Arentze, T., Timmermans, H., Janssens, D., & Wets, G. (2011). The effects of social networks on choice set dynamics: Results of numerical simulations using an agent-based approach. *Transportation Research Part A: Policy and Practice, 45*(4), 310–322.

Hollstein, B. (2011). Qualitative approaches. *The SAGE Handbook of Social Network Analysis*, 404–416.

Holme, P., & Saramäki, J. (2012). Temporal networks. *Physics Reports, 519*(3), 97–125.

Huang, L., & Knight, A. P. (2017). Resources and relationships in entrepreneurship: An exchange theory of the development and effects of the entrepreneur-investor relationship. *Academy of Management Review, 42*(1), 80–102.

Illenberger, J., Flöteröd, G., Kowald, M., & Nagel, K. (2009). A model for spatially embedded social networks. Paper presented at the 12th International Conference on Travel Behavior Research (IATBR), Jaipur.

Jackson, M. O. (2007). The study of social networks in economics. In J. E. Rauch (Ed.), *The missing links: Formation and decay of economic networks*. New York: Russell Sage Foundation.

Kim, J., Rasouli, S., & Timmermans, H. J. P. (2014a). Expanding scope of hybrid choice models allowing for mixture of social influences and latent attitudes: Application to intended purchase of electric cars. *Transportation Research Part A: Policy and Practice*, *69*, 71–85.

Kim, J., Rasouli, S., & Timmermans, H. J. P. (2014b). Hybrid choice models: Principles and recent progress incorporating social influence and nonlinear utility functions. *Procedia Environmental Sciences*, *22*, 20–34.

Kim, J., Rasouli, S., & Timmermans, H. J. P. (2018). Social networks, social influence and activity-travel behaviour: A review of models and empirical evidence. *Transport Reviews*, *38*(4), 499–523.

Kowald, M., & Axhausen, K. W. (2012). Focusing on connected personal leisure networks: Selected results from a snowball sample. *Environment and Planning-Part A*, *44*(5), 1085.

Kowald, M., & Axhausen, K. W. (Eds.). (2015). *Social networks and travel behaviour*. Surrey: Ashgate Publishing Ltd.

Kowald, M., van den Berg, P., Frei, A., Carrasco, J. A., Arentze, T., Axhausen, K. . . . Wellman, B. (2013). Distance patterns of personal networks in four countries: A comparative study. *Journal of Transport Geography*, *31*, 236–248.

Lovrić, M., Da Re, R., Vidale, E., Pettenella, D., & Mavsar, R. (2018). Submission of an original research paper: Social network analysis as a tool for the analysis of international trade of wood and non-wood forest products. *Forest Policy and Economics*, *86*, 45–66.

Mayer, A. (2009). Online social networks in economics. *Decision Support Systems*, *47*(3), 169–184.

McCowan, B., Beisner, B. A., Capitanio, J. P., Jackson, M. E., Cameron, A. N., Seil, S. . . . Fushing, H. (2011). Network stability is a balancing act of personality, power, and conflict dynamics in rhesus macaque societies. *PLoS One*, *6*(8), e22350.

Miguéns, J. I. L., & Mendes, J. F. F. (2008). Travel and tourism: Into a complex network. *Physica A: Statistical Mechanics and Its Applications*, *387*(12), 2963–2971.

Mitrović, M., & Tadić, B. (2012). Emergence and structure of cybercommunities. In *Handbook of optimization in complex networks* (pp. 209–227). Boston, MA: Springer.

Molin, E., Arentze, T., & Timmermans, H. J. P. (2008). Social activities and travel demand: Model-based analysis of social network data. *Transportation Research Record: Journal of the Transportation Research Board*, *2082*, 168–175.

Mollick, E. (2014). The dynamics of crowdfunding: An exploratory study. *Journal of Business Venturing*, *29*(1), 1–16.

Montgomery, J. D. (1991). Social networks and labor-market outcomes: Toward an economic analysis. *The American Economic Review*, *81*(5), 1408–1418.

Newman, M. E. (2002). Assortative mixing in networks. *Physical Review Letters*, *89*(20), 208701.

Orsini, C., Dankulov, M. M., Colomer-de-Simón, P., Jamakovic, A., Mahadevan, P., Vahdat, A. . . . Krioukov, D. (2015). Quantifying randomness in real networks. *Nature Communications*, *6*, 8627.

Páez, A., & Scott, D. M. (2007). Social influence on travel behavior: A simulation example of the decision to telecommute. *Environment and Planning A*, *39*(3), 647–665.

Páez, A., Scott, D. M., & Volz, E. (2008). A discrete-choice approach to modeling social influence on individual decision making. *Environment and Planning B: Planning and Design, 35*(6), 1055–1069.

Palchykov, V., Mitrović, M., Jo, H. H., Saramäki, J., & Pan, R. K. (2014). Inferring human mobility using communication patterns. *Scientific Reports, 4*, 6174.

Palla, G., Barabási, A. L., & Vicsek, T. (2007). Quantifying social group evolution. *Nature, 446*(7136), 664.

Park, J. Y., Im, I., & Sung, C. S. (2017). Is social networking a waste of time? The impact of social network and knowledge characteristics on job performance. *Knowledge Management Research & Practice, 15*(4), 560–571.

Pastor-Satorras, R., Castellano, C., Van Mieghem, P., & Vespignani, A. (2015). Epidemic processes in complex networks. *Reviews of Modern Physics, 87*(3), 925.

Polzin, F., Toxopeus, H., & Stam, E. (2018). The wisdom of the crowd in funding: Information heterogeneity and social networks of crowdfunders. *Small Business Economics, 50*(2), 251–273.

Rasouli, S., & Timmermans, H. (2014). Activity-based models of travel demand: Promises, progress and prospects. *International Journal of Urban Sciences, 18*(1), 31–60.

Ronald, N., Dignum, V., Jonker, C., Arentze, T., & Timmermans, H. J. P. (2012). On the engineering of agent-based simulations of social activities with social networks. *Information and Software Technology, 54*(6), 625–638.

Sadri, A. M., Ukkusuri, S. V., & Gladwin, H. (2017). Modeling joint evacuation decisions in social networks: The case of Hurricane Sandy. *Journal of Choice Modelling, 25*, 50–60.

Scott, N., Baggio, R., & Cooper, C. (2008). *Network analysis and tourism: From theory to practice*. Clevedon: Channel View Publications.

Sen, P., & Chakrabarti, B. K. (2014). *Sociophysics: An introduction*. Oxford: Oxford University Press.

Shane, S., & Cable, D. (2002). Network ties, reputation, and the financing of new ventures. *Management Science, 48*(3), 364–381.

Sharmeen, F. (2015). *Dynamics of social networks and activity travel behaviour*. Doctoral Dissertation, Technische Universiteit Eindhoven, Eindhoven.

Sharmeen, F., Arentze, T., & Timmermans, H. J. P. (2013). Incorporating time dynamics in activity travel behavior model: A path analysis of changes in activity and travel time allocation in response to life-cycle events. *Transportation Research Record: Journal of the Transportation Research Board, 2382*, 54–62.

Sharmeen, F., Arentze, T., & Timmermans, H. J. P. (2014a). An analysis of the dynamics of activity and travel needs in response to social network evolution and life-cycle events: A structural equation model. *Transportation Research Part A: Policy and Practice, 59*, 159–171.

Sharmeen, F., Arentze, T., & Timmermans, H. J. P. (2014b). Dynamics of face-to-face social interaction frequency: Role of accessibility, urbanization, changes in geographical distance and path dependence. *Journal of Transport Geography, 34*, 211–220.

Sharmeen, F., Arentze, T., & Timmermans, H. J. P. (2015a). Dynamic social networks and travel. In M. Kowald & K. Axhausen (Eds.), *Social networks and travel behaviour* (pp. 203–218). London: Routledge.

Sharmeen, F., Arentze, T., & Timmermans, H. J. P. (2015b). Predicting the evolution of social networks with life cycle events. *Transportation, 42*(5), 733–751.

Sharmeen, F., Chávez, Ó., Carrasco, J. A., Arentze, T., & Tudela, A. (2016). A modelling population-wide personal network dynamics using a two-wave data collection method and an origin-destination survey. In *TRB 95th Annual Meeting Compendium of Papers*. Washington, DC: Transportation Research Board.

Sharmeen, F., & Sivakumar, A. (2017). Why care about social networks in travel demand forecasting? Testing the predictive power of social attributes in modeling discretionary trip frequencies. In *TRB 96th Annual Meeting Compendium of Papers*. Washington, DC: Transportation Research Board.

Sharmeen, F., & Timmermans, H. J. P. (2014). Walking down the habitual lane: Analyzing path dependence effects of mode choice for social trips. *Journal of Transport Geography*, *39*, 222–227.

Sheller, M., & Urry, J. (2006). The new mobilities paradigm. *Environment and Planning A*, *38*(2), 207–226.

Silvis, J., Niemeier, D., & D'Souza, R. (2006). Social networks and travel behavior: Report from an integrated travel diary. Paper presented at the 11th International Conference on Travel Behavior Research, Kyoto.

Tibély, G., Onnela, J. P., Saramäki, J., Kaski, K., & Kertész, J. (2006). Spectrum, intensity and coherence in weighted networks of a financial market. *Physica A: Statistical Mechanics and Its Applications*, *370*(1), 145–150.

Timur, S., & Getz, D. (2008). A network perspective on managing stakeholders for sustainable urban tourism. *International Journal of Contemporary Hospitality Management*, *20*(4), 445–461.

Travers, J., & Milgram, S. (1967). The small world problem. *Psychology Today*, *1*(1), 61–67.

Urry, J. (2003). Social networks, travel and talk. *The British Journal of Sociology*, *54*(2), 155–175.

Urry, J. (2004). Small worlds and the new 'social physics'. *Global Networks*, *4*(2), 109–130.

van den Berg, P., Arentze, T., & Timmermans, H. J. P. (2008). Social networks, ICT use and activity-travel patterns. In *Data collection and first analyses*. Paper presented at the 9th International Conference on Design & Decision Support Systems in Architecture and Urban Planning, The Netherlands.

van den Berg, P., Arentze, T., & Timmermans, H. J. P. (2009). Size and composition of ego-centered social networks and their effect on geographic distance and contact frequency. *Transportation Research Record: Journal of the Transportation Research Board*, *2135*, 1–9.

van den Berg, P., Arentze, T., & Timmermans, H. J. P. (2010). Factors influencing the planning of social activities: Empirical analysis of data from social interaction diaries. *Transportation Research Record: Journal of the Transportation Research Board*, *2157*, 63–70.

van den Berg, P., Arentze, T., & Timmermans, H. J. P. (2011). Estimating social travel demand of senior citizens in the Netherlands. *Journal of Transport Geography*, *19*(2), 323–331.

van den Berg, P., Arentze, T., & Timmermans, H. J. P. (2012). Involvement in clubs or voluntary associations, social networks and activity generation: A path analysis. *Transportation*, *39*(4), 843–856.

van den Berg, P., Arentze, T., & Timmermans, H. J. P. (2013). A path analysis of social networks, telecommunication and social activity – travel patterns. *Transportation Research Part C: Emerging Technologies*, *26*, 256–268.

van den Berg, P., Weijs-Perrée, M., & Arentze, T. (2017). Dynamics in social activity-travel patterns: Analyzing the role of life-cycle events and path dependence in face-to-face and ICT-mediated social interactions. *Research in Transportation Economics*, *68*, 29–37.

Wang, G., & Wang, Y. (2018). Herding, social network and volatility. *Economic Modelling*, *68*, 74–81.

Wasserman, S., & Faust, K. (1994). *Social network analysis: Methods and applications* (Vol. 8). Cambridge: Cambridge University Press.

Watts, D. J., & Strogatz, S. H. (1998). Collective dynamics of 'small-world' networks. *Nature*, *393*(6684), 440.

Wellman, B., Wong, R. Y.-l., Tindall, D., & Nazer, N. (1997). A decade of network change: Turnover, persistence and stability in personal communities. *Social Networks*, *19*(1), 27–50.

2 How to define social network in the context of mobilities

Bridgette Wessels, Sven Kesselring and Pnina O. Plaut

Introduction

Social activities steaming from social relations are one of the main motivators for the use of transport systems. A social network is a social structure, based upon group members and the relations among them. Social network studies are concerned with the structure of sociocultural systems, such as the number of contacts and the levels of communication among different members of a certain social group. The past decade has witnessed rapid communication developments, which had major social impacts. The use of new Information and Communication Technologies accelerated the shift from social groups that were defined through a specific location (e.g. residential neighbourhood or workplace) to individually based social networks. This shift, coined by Wellman (2001, 2002) as "networked individualism", is a stage in which mobile, high-speed telecommunication allows for personalized networks and "person-to-person" social ties. These new social networks are associated with several changes, when comparing to the past 50 years, including people having a larger set of active contacts today than in the past, with wider spatial distribution, the contacts spread across more social networks than in the past and typical social networks less coherent; that is fewer people share multiple affiliations today than in the past, and leisure travel is increasing (Axhausen, **2005**). All of the preceding have an impact on travel and mobility within the urban realm.

In this chapter, we present social networks in the context of mobilities. We argue that traditional approaches to travel and travel behaviour based on ego-centred perspectives do not fully address the way in which travel is embedded within social networks and connections. We argue that senses of mobility are created from the ways in which people meaningfully organize their lives. Mobility is generated through social networks, both personal and impersonal or as Urry (2003) puts it "social networks, travel and talk" are closely and indissolubly intertwined.

Egocentred approaches to the analysis of social networks and travel behaviour

Research on the impact of Information and Communication Technology (ICT) social networks on travel behaviour is still in its infancy (about a decade old).

Facebook, the most popular and influential social network, is just **13** years old, and smartphones are even younger than that. Research in the field begun between 2003 and 2006. There were four research groups leading the study into relations between social networks analysis and activity-travel modelling approach (Table 2.1). As pioneers in this field, they had to deal with basic questions, such

Table 2.1 Classification and summary of pioneering studies

	Group 1: Canada/Chile Wellman, Carrasco and Hogan (NetLab)[*]	Group 2: Switzerland/ Germany/UK Axhausen, Frei, Kowald, Ohnmacht, Urry, Larsen[**]	Group 3: U.S Silvis, Niemeier, D'Souza[***]	Group 4: Netherlands Timmermans, Van den Berg, Arentze[****]
place/ year	Toronto, Canada and Concepción, Chile	Lancaster (UK), Zürich and Berlin	Davis, California	Eindhoven region
	2004 – 2005 and 2008 – 2009	2004–2005, 2005–2006, 2009 – 2011	2005	2008
Number of respondents	350 in Canada and 120 in Chile	24 (Lancaster), 30 (Zürich and Berlin) and 307+743 (Zürich)	24	747
Approach/ Method	Egocentric Surveys and interviews	Egocentric Questionnaires, interviews, and social interaction diary	Egocentric Interviews	Egocentric Social interaction diary and questionnaires
Network boundary definition	Weak/Strong tie	Weak/Strong tie	Weak/Strong tie	Weak/Strong tie
Type of findings	Graphic Sociogram, correlations between socio-economic characteristics and travel behaviour	Mapping mobility biography travels Mapping "Confidence ellipse" (= activity spaces)	Defining two types of social travels based on social interactions	Activity-based models, path regression model

[*] Carrasco, Hogan, Wellman, & Miller, 2008; Carrasco & Cid-Aguayo, 2012.
[**] Axhausen, 2005, 2008; Larsen, Urry, & Axhausen, 2006; Ohnmacht, 2006; Frei & Axhausen, 2007; Axhausen & Frei, 2008; Kowald, Frei, Hackney, Illenberger, & Axhausen, 2009; Kowald & Axhausen, 2010.
[***] Silvis, Niemeier, & D'Souza, 2006.
[****] Van den Berg, Arentze, & Timmermans, 2009, 2010.

as Which social network parameters are important for travel behaviour study? Which phenomena should be studied? and What methods are suitable? We are still facing these questions today.

The studies used surveys, travel diaries and interviews for data collection. An agent-based analysis was performed. They usually approach the participants for the survey through snowball sampling or neighbourhood-based sampling. The general practice shared by all previous research is that travel behaviour, though influenced by social networks, is being studied on an individual basis – that is by using the "Egocentric Analysis" approach. A recent literature review on models and empirical evidence confirms the dominance of the egocentric approach in social networks and activity-travel behaviour studies (Kim, Rasouli, & Timmermans, 2018).

Analysis of networks through an egocentric approach takes as its starting point one agent (ego node) and examines its connections with other members (alters) in his or her network. It is usually used for first-order ego networks (Wellman & Marin, 2011). The analysis is focused on affiliation ties. The relations measured within the analysis is divided into several types of inquiries following Wasserman and Faust (1994), including the following:

- Individual sentiments regarding the positive or negative affect of one person for another, including friendship, liking, respect and so on
- Exchange or transfer of resources, including material support or specific forms of social support (such as lending money)
- Exchange or transfer of non-material resources, such as received or sent messages, giving or receiving advice and providing important information
- Interactions which involve the physical presence of two parties, such as sitting next to each other, attending the same party, visiting a person's home, hugging, conversing and so on
- Formal roles such as boss/employee, teacher/student or doctor/patient
- Kinship relations such as parent/child, marriage, close friends or acquaintances

The relations happening between two people may be reciprocal or may be only a tie in one direction. They are usually evaluated as a dichotomy between "strong" or "weak" ties. There was also an attempt to define the strength of contacts on a more continuous scale introducing the concept of "in-betweens" (Huszti, David, & Vajda, 2013). Others address the dynamics of social networks based on the fact that networks are not fixed over time (Sharmeen, 2015). The dominant questioning format for egocentric analysis is the name generator. Its aim is to identify the strength of ties, as well as its boundaries. Usually it follows with a name interpreter in which details are given per each contact. Name generators could be divided into three groups (Bidart & Charbonneau, 2011):

1 Generators that are based on interactions: identifying the persons encountered the most frequently, those people we meet or those we communicate with by telephone or ICT modes over a given period (e.g. last week, month or

year). A typical question within the name-generator tool might be, "By think-ing of all the people with whom you are in contact, please name those with whom you are most often in contact?"

2   Generators that are based on the importance of certain links: identifying the persons to whom we feel the closest, best friends; "persons of greatest importance to the ego" (Wellman, 1979); or those with the greatest impact on one's attitudes, behaviour and welfare (McCallister & Fischer, 1978). A typi-cal question might be, "From time to time, most people discuss important personal matters with other people. Recalling the last 6 months, who are the people with whom you have discussed important personal matters?" (Usually the respondent is asked to name five such contacts).

3   Generators based on exchange: identifying the persons who are likely to procure various resources for us (help, information, emotional support, and advice). One known example is Fischer's (1982) name generator, which questions and addresses various notions of support. The respondents were asked to name the following: those people whom respondents would ask to look after their homes when they go out of town; those who had helped with tasks around the house in the previous three months; those people whom respondents would talk with about how they do their jobs; those with whom respondents do various social activities, such as sharing a meal, visiting, going out socially and so on; those with whom respondents used to date seri-ously or be fiancé or fiancée; those with whom respondents talk with about personal matters of concern; those people that respondents rely on for advice about important decisions; or those that respondents would ask for a sizeable loan.

Once a list has been made, respondents are asked to fill a name interpreter ques-tionnaire, which gathers information on each of the listed contacts. In general, these questions seek to identify the sociodemographic characteristics of each con-tact, the relationship between the respondent and the contact and the relationships between contacts. Cognitive data questionnaire is another widely used question-ing format for egocentric analysis, focusing on perceived relations within network members. In this case, the respondents (egos) are asked to fill in sociograms, indicating their perceived relations with other actors within their network (i.e. close friends, acquaintances, spouse/family members, other relatives, work asso-ciates, etc.). This tool is helpful within the egocentric studies, in which boundaries are more fluid and relations are observed via one participant. Examples of name generators that are used in studies on social networks and travel behavior are "very-close people: people with whom you discuss important matters, or who you regularly keep in touch with, or who are there for you if you need help" and "somewhat-close people: people who are more than just casual acquaintances, but not very close" (Carrasco et al., 2008). The questions can also be developed to elicit specific alters associated with social activities, such as "people with whom the respondents spend leisure time" (Frei & Axhausen, 2007) and "people with whom you make plans to spend free time" (Kowald & Axhausen, 2010).

Most studies conducted to date have used the traditional transport activity-based modelling and variables, in particular travel distances, frequencies and mode of travel. These studies focused on leisure trips as yet one more purpose, distinguishing it from travel to work or travel to shopping. The studies resulted in drawing "confidence ellipse", showing distances and activity space of an average person (e.g., Schönfelder & Axhausen, 2003; Frei & Axhausen, 2007; Ohnmacht, 2006; Silvis et al., 2006), or in statistical technical correlations among travel variables (such as social travel distance, mode of travel, number of trips) and social network characteristics (such as strong/weak ties between members), controlling for personal and household characteristics. (Axhausen, 2008; Kowald & Axhausen, 2016; Kowald et al., 2013). Thus, we see that the egocentric approach to characterize the social network ties coincides (or is in congruence) with the traditional travel behaviour approach and activity-based modelling in transportation studies. Both focus on the individual and his or her personal social ties correlating it with his personal travel behaviour. These analyses usually consisted of four elements, aiming to capture the nature of the "activity space" from the ego point of view:

- Social density: Capturing how many connections are made between persons at a given network (range between "tight" to "loose" networks) and what type of communication is used (face-to-face, cellphone, Facebook, etc.)
- Strength of connection: The distinction between weak and strong ties (which is a definition for closeness between members in the network)
- Mobility and travel behaviour: What modes of transport are taken, purpose and frequency of the travel
- Spatial density: Capturing the physical location, distances and type (dense/ sparse) of where social networks take place

This was a short summary of the egocentric approach as it has been applied in transport research so far.

People, networks and mobility

This section discusses how the meaning of travel and how travel is embedded within social networks. Travel involves a range of activities and people define travel in terms of the purpose of the travel, the experience of travel and its social and personal meaning. There is an increase in levels of personal mobility seen in increasing trends of longer travel distances and more frequent local and regional travel (Frändberg & Vihelmson, 2003). Examples of personal mobility include a short trip into the city centre, a commute to work, business travel, long-distance travel to go on holiday, the school run in the car and a community bus to go to the day-care centre. Each of these examples is meaningful for those travelling, and they shape the meaning of mobility in contemporary society. Part of the way meanings of travel develop is through the usage of travel time. For example, travel on trains, buses and planes provide people with time to undertake work, sleep, rest, read and window gaze (Lyons, Jain, & Holley, 2007). Another aspect in the utilization of travel time is ICT (Thulin & Vilhelmson, 2008). Gustavson (2012) argues that travellers may well consider not just the length of the travel time but, rather, what they can do while travelling using ICT. By combining the meaningfulness of travel, its purposes and networked configuration travel are embedded in mobilities, what we mean by this is the ways in which transport resources are used to undertake different types of travel that are enacted through social networks.

In this section, we argue that the ways in which people organize their lives, with whom and with what they connect with, shapes their mobilities. Mobility is generated through social networks that have both personal and impersonal connections (see Urry, 2003). Mobility in contemporary society involves drawing on a range of transport resources and configuring how to use those in relation to the meanings and purposes of travel. The specific characteristics of networks vary depending on the social and personal context of each type of mobility and travel. This section takes forward our argument that travel in contemporary society is shaped by how people develop networks through the meanings and the connections they have with other people as well as with events, organizations and places.

Organizing social life: networks and meaning

Social life is organized through time and space through a range of social practices (Giddens, 1984). Although there is debate about how social life is structured, it is nonetheless recognized that people act and organize their activities using the resources they have within certain structural frameworks. To address the concern of this chapter, we need to address two aspects of the last point, which are (a) the forms of organizing social lives and (b) the meanings that underpin that organization. This means paying attention to the patterns of how people organize and live their lives and why people organize their lives in particular ways, thus addressing the meanings that underpin decisions about what to do, who to meet, where to go and so on. Both the patterning of activities and the meanings of activity are important because they inform and shape mobilities (Urry, 2000).

In broad terms, social networks refer to a social structure that forms through a set of social actors who interact within dyadic ties and a range of broader social interactions. Some of those ties are stronger than others, and the shape of the network will vary in relation to the way actors' ties and relations configure. There is a long history in social network theory. Tönnies (1957), for example, writes that social groups exist through (a) personal and direct social ties they may link individuals with shared values and beliefs, often felt of as Gemeinschaft, and/or (b) links that are more impersonal, formal and instrumental, often felt as Gesellschaft. Simmel (1964), writing a little later, developed ideas about webs of group affiliations, identifying networks and networking characteristics, such as effects of the size of networks on interaction and interaction in loosely knit networks as well as in more tightly defined groups. In the development and use of digital technologies and services, and a turn to networked society, the idea of a social network is developed further. More recent scholars in the field include Rainie and Wellman (2012) who focus on networks in the digital age.

The development of a digitally supported network society (Castells, 2001) and social and cultural changes of late modernity have institutionalized social networks more deeply into society (Wessels, 2014). Wellman and Haythornthwaite (2002) suggest that society is changing from one of the bounded communities to numerous individualized, fragmented, personalized communities and that digital services underpin a networked society. Although uneven, in general terms, the organization of social life has moved to networks made up of resources, information/data, goods, services, organizations and people. These resources are to varying extents embedded in people's social networks, and they can access and mobilize these resources through the ties in the network. The resources available through a person's social network form that person's social capital. Social capital is defined as the resources people have in their respective social networks, which they can mobilize or use (Lin, 2001). Many scholars argue that social capital is networked-based (Bourdieu, 1980; Bourdieu, 1983/1986; Coleman, 1990; Putnam, 2000), and mobility networks are one example of social capital. These networks involve modes of transport, information and data, transport goods, services, organizations, staff and passengers/travellers, as well as transport resources, including knowledge of travel and material resources, such as money to pay for travel. In terms of mobilities, access to travel and transport resources are important in supporting individuals in developing social capital as well as supporting them to mobilize social capital.

The purposes and experiences of travel and the personal and social aspects of travel create the meaning of mobilities. To address meaning and meaningful action requires recognizing human agency. One of the founding scholars of sociology, Max Weber, developed a theoretical approach to address meaning in social action. Weber (1922) argues that meaning is attached to action and that individual and group actions are shaped and activated through meaning. This raises questions about the way in which meaning is attached to action, and here, insights from phenomenology are helpful. Schutz (1962, 1971) building on Weber's idea that human action is meaningful makes human agency and experience theoretical priorities. His point of departure is the individual's own definition of the situation. Individuals in defining his or her situations draw on a common stock of knowledge, which is made up of social conceptions that relate to types of things, material or symbolic. It is through these "typifications" that individuals develop routines and definitions of situations within their own social worlds.

Blumer (1969) developed these ideas about meaning and action, which resulted in the theory of Symbolic Interactionism (SI) (Blumer, 1969). The key aspects of SI are

- human beings act towards things on the basis of the meanings that things have for them,
- that meanings of things arise out of the social interaction that people have with each other and
- that the meanings of things are handled in and modified through an interpretive process used by the person in dealing with things they encounter.

What these approaches show is that the human element of a social network is characterized by the way that people make sense of their situations and undertake meaningful action. These interpretations, actions and meanings are embedded within social contexts and networks. They can also develop structural properties such as routines, repertoires and rituals. These routines, repertoires and rituals, and other non-routine meaningful action are embedded into networking logics of mobilities.

Mobilities are therefore networked and meaningful; they are crafted out of the resources of various types of mobilities, social networks and meaningful action. The history of transport, travel and mobility demonstrates that the characteristics of mobility changes in historical time. To address current mobilities means understanding the characteristics of contemporary social networks, new types of resources and knowledge.

Networked organizations, networked individualism and mobilities

Castells (2001) argues that the way the economy and society has adapted the use of the Internet and World Wide Web (WWW) has ushered in a networked society. What he means by this is that the organizational form of society and the economy is based on a network enabled by a digital infrastructure. In economic terms, business is now organized through a networking logic of production, trade and distribution. The technology does not determine this change; rather, the neo-liberal globalized economy could utilize the networked design of the Internet and WWW. The networking logic extends into the ways organizations operate. Transport organizations are part of this change as well as other organizations.

These sorts of changes also extend into the social realm, with Wellman and Haythornthwaite (2002) arguing that one of the characteristics of current social change is the emergence of a changing relationship between the individual and society, which they call "networked individualism". This does not refer to any specific individual but, instead, needs to be seen as a social form. What Wellman and Haythornthwaite (2002) mean by "networked individualism" is that the form involves the ways in which individuals create social networks by selecting other individuals to join their network often using digital communication to do so. These networks can be dispersed across time and space, and individuals manage them (Rainie & Wellman, 2012). This transition is seen as being from groups with defined hierarchies to networks involving the mobility of people and goods. As noted earlier, Wellman and Haythornthwaite (2002) suggest that society is changing from one of the bounded communities to numerous individualized, fragmented, personalized communities and that digital technology facilitates this development of networked society.

Although not reduced to digital technologies of the Internet and WWW, networked individualism draws on digitally mediated communication. This includes personalized, portable, ubiquitous connectivity and wireless mobility, which facilitate networked individualism as a basis for social and cultural life (Rainie & Wellman, 2012). In this context, the individual has a significant role to play in networks because he or she is the primary unit of connectivity and is active in generating communication and action within a network, including networks of mobilities. There are connections and ties with of the mobilities resources, such as modes of transport, travel information, transport organizations and others such as travellers in a mobilities network. These types of resources are publicly available, and nonetheless, human agency is involved in these networks by creating meaningful links and generating and interpreting particular mobilities and situations and instances of mobility (Silverstone, 2005; Berker, Hartmann, Punie, & Ward, 2005).

Grounding and practising mobilities in social networks

Organizational networks, networked individualism, institutions and digital communication feature in, and are part of, mobility networks. They also combine to make mobility meaningful and here knowledge and information feature in that meaning-making. The way in which everyday life and organizational life are based on digitally enabled social networks and mobility networks raises questions about particular sites or nodes in a network. Here, social institutions and organizations feature and play a particular role in social networks. Research shows that social institutions such as the household are significant in shaping technology and everyday life (Silverstone, 2005). The household forms a node as a site that distributes resources, takes up and adapts innovations and grounds networked individualism (Wessels, 2010, 2014).

Individuals, organizations, modes of transport, transport services and information and knowledge are part of mobilities networks. Digital technology has a distinctive role in the production of knowledge because it has the capacity to create knowledge from a range of sources and networks. Mobilities, when embedded within social networks, involve technological, as well as human, systems. Technological systems are intelligent in that they link, combine and compute data to create new knowledge for transport systems, services and production (Lash, 1999; Lyotard, 1984), and humans, passengers, transport services workers create and understand information to create knowledge within cultural frameworks that frame mobilities (Geertz, 1973).

The practices of mobilities are therefore both contextually grounded in people, transport resources, places and institutions and contextually networked in flows of mobilities, in the networks of modes of transport, routes and timetables and places. So although transport and travel are networked, human agency features in the understanding and interpretive and active dimensions of social networks and mobilities. Social networks and mobilities are part of the networking logic of the digital age.

Social networks and mobilities

The characteristics of a digital age point to the emergence of new social forms – first, the overall networking logic that links different aspects of organizational and individual life. The organization of social life is based in a network that distributes and organizes the resources of mobility (Castells, 2001). Networked individualism as a social form emerges from, and is located in, the needs and desires of individuals to organize mobilities within their social networks. The network facilitates them to construct meaningful mobilities by crafting time, space and a range of resources through digital technology and a myriad of social practices and knowledge. Second, networked individualism is the form through which individuals select significant others, places, travel and transport modes in creating their mobilities. These relationships show how individuals seek to manage space, time and mobilities within networked relations. The role of digital technology is influential in facilitating networks and mobilities, but social networks are not reducible to the technology. Rather, human agency is important within these networks because the meaning of the mobility, its value for people and the volition to be part of a range of mobilities is generated through people's agency. However, the social network is more than agency; it is also constituted through the resources of mobilities and mobility networks. These involve modes of transport, information and data, transport goods, services, organizations, staff and passengers/travellers, as well as transport resources, including knowledge of travel and material resources such as money to pay for travel.

All this has to be assessed and reflected against the background of a currently fastly transforming "system of automobility" (Urry, 2004). If we say "system of automobility", we consider this as a comprehensive and highly complex socio-material networked structure in which the car plays the role of the iconic modern mode of transport (Mom, 2015; Sachs, 1992; Freudendal-Pedersen & Kesselring, 2018a). But more than this, the automobile is essentially pre-structuring and dominating the way how individuals in modern societies live, consume, produce and interact (Urry, 2004). For more than 100 years, economic production and reproduction have been configured around the idea that people and goods, resources and even information (in the form of books, magazines, files, etc.) need to be physically transported by vehicles. Other mobility systems, such as public transport, seafaring and aeromobilities, have been developed and organized around the car as the key connector in modern societies (Cwerner, Kesselring, & Urry, 2009; Cidell, 2017; Dennis & Urry, 2009).

This system is in deep-going and most likely irreversible transformations (Geels, 2012; Dennis, 2013; Freudendal-Pedersen, Kesselring, & Servou, 2019). The current debate on diesel cars in Europe and the United States, the so-called Dieselgate (Hilgenberg, 2017), is just an indicator for an evolution away from the centrality of the automobile and towards a new digital age of and a system of multi-mobility where networked technologies in combination with intelligent transport systems will replace the current monocentric technostructure mobilities (see Canzler & Knie, 2016; Diez, 2018). These days diesel cars play the role of a sort of boogieman, even if there are good reasons for going beyond this technology. It is a sort of disenchantment of the concept of the driver–owner–car (namely the diesel car) and the rise of Mobility-as-a-Service (MaaS; Audouin & Finger,

2018) as a contemporary form of mobility and transport and as a new foundation for social networks and connectivities.

Platforms such as GoMore, the Danish car-sharing and carpooling company, with 2.4 million users, stands for a new networked form of mobility services which transgresses the idea of the necessity to own a car to be a full member of society.

Car producers, such as BMW, push forward what they call 'connected life' and for what they have different perceptions and concepts in mind. There is still the artefact of the car in the centre of their business model, but the car has already become a sort, if a virtualized and mediatized, product, a material object which is embedded and intertwined with virtual and social networks and shaped through media technologies of information, infotainment and entertainment. The new generation of cars in the premium segment increasingly appears as high-end computers on four wheels. Not as much the steering wheel and its environment are in the centre of the mobility device but, rather, the computer screen which enables activating music streaming, opening doors, "steering" heat, air, music, videos, TV and so forth. Beyond its functionality as a control centre and an entertainment console, the screen provides an entry point to all sorts of social media and networks. Connected with the smartphone, it enables the driver for all sorts of communications and interactions while driving or waiting in congestion and so forth.

In the new world of multi-mobility based on (shared) cars, bikes, scooters, public transport, shared rides and so forth (see Freudendal-Pedersen & Kesselring, 2018b) mobility and transport are already happen to become a mediatized, digitalized and socially networked experience. For many people, almost every sphere of private and professional life is connected and linked by technologies, often without the users' awareness. There is a "technological unconsciousness" (Elliott & Urry, 2010) at work on which travel and communication rest and where the users of transportation facilities, mobile phone apps and websites, among others, forget – or do not even know or care – the constant flows of data that are collected throughout their travels. People are not only flying through "code/space" (Dodge & Kitchin, 2004), maybe even better formulated as coded space; they are, rather, travelling through it wherever they go. Be it on public transport or when using one of the new biking schemes where they effectively "pay" with data (Spinney & Lin, 2018). "Movement space" (Thrift, 2004), the physical and social media spaces where algorithms are permanently checking and collecting changes in space (if not actively switched off on mobile devices), is iconic and paradigmatic for the current level of modernization. The "new culture of AI [artificial intelligence]" (Elliott, 2018) must be considered not only to be deep-going and transformative for social life, community, interaction, individualization and mobility; he literally calls it a "technological tsunami" hitting the social lives of modern people. By detecting individual preferences, connections and visits in virtual and physical space bots and robots are collecting big data volumes from which providers can reconstruct movement and communication patterns and their influences on travel and the use of space and infrastructures.

Through the massive rise of artificial intelligence in all sorts of social, communicative, business and other activities the boundaries between social interaction, networking and technologically-based forms of it become fluid (Turkle, 2011; Elliott, 2016). The current discussion on automated driving and autonomous vehicles and the often-high-flying hopes of industry, planners and politics shed light on the possible transformations of social networks and travel behaviour.

Conclusion

As we have shown in the first and the second part of the chapter, in the digital age we can expect even denser and more dynamic forms of social networks, be they mediatized or quasi–"old school" based on face-to-face contacts and interactions. Artificial intelligence increasingly will be part of the process of the social construction of reality (Couldry & Hepp, 2017; Morley, 2017). While in Simmel's "old days" the individual was the actant and the subject, actively building social ties, connections, close intimate relations and who invited people into its social circles (Simmel, 1923, 1964, 1971), today, technological agents are being part of offering possible connections. Facebook's yearly reminders to birthdays and its presentation of pictures from the individual's gallery are just simple forms. Together with its never-ending presentation of possible contacts or "friends", it actively interferes into the configurations and morphologies of social networks. In other words, new social media become essential elements in building and maintaining social worlds and furthermore the individual's "opportunity spaces" (Winnicott, 1997). Following Wellman's and others' findings on the relationships between social networks and direct face-to-face interactions, the intensification of "communication work" often leads to physical meetings and stable travel connections (Larsen et al., 2006; Kesselring, 2006; Freudendal-Pedersen & Kesselring, 2018a). German philosopher Hermann Lübbe even claims that the more people, institutions and networks communicate, the more they travel (Lübbe, 1993). John Urry's observations of the new "culture of meetingness" directly feeds into this. For Urry, the need for travel and proximity, for meeting face-to-face friends, relatives, co-workers and customers will probably not end even in a society with highly sophisticated and developed infrastructures for virtual and digital communication and interaction (see Urry, 2002). But even when face-to-face interaction remains an ontological constant in human life, the future of mobility and travel will be digitalized and transformed through all sorts of smart technologies. The underlying artificial intelligence in social media networks has a strong potential to permeate modern lives to a very high degree. In this chapter, we wanted to indicate the urgency and progress in research on social networks and travel. And this is mainly the case because, in line with Maarten Hajer's call for a smart urbanism instead of a blind faith in smart technologies (Hajer & Dassen, 2014), we see the significant need of further research in the intended and unintended consequences of the coming together of travel and social network activities.

References

Audouin, M., & Finger, M. (2018). Empower or thwart? Insights from Vienna and Helsinki regarding the role of public authorities in the development of MaaS schemes. *Transportation Research Procedia*, 1–11. Retrieved from www.sciencedirect.com

Axhausen, K. W. (2005). Social networks and travel: Some hypotheses. *Social Dimensions of Sustainable Transport: Transatlantic Perspectives*, 90–108.

Axhausen, K. W. (2008). Social networks, mobility biographies, and travel: Survey challenges. *Environment and Planning B: Planning and Design*, 35, 981–996.

Axhausen, K. W., & Frei, A. (2008, January). Contacts in a shrunken world. Paper presented at the 86th Annual Meeting of the Transportation Research Board, Washington, DC.

Berker, T., Hartmann, M., Punie, Y., & Ward, K. (Eds.). (2005). *Domestication of media and technologies*. Maidenhead: Open University Press.

Bidart, C., & Charbonneau, J. (2011). How to generate personal networks: Issues and tools for a sociological perspective. *Field Methods*, 23(3), 266–286. https://doi.org/10.1177/1525822X11408513

Blumer, H. (1969). *Symbolic interactionism: Perspective and method*. Englewood Cliffs, NJ: Prentice-Hall.

Bourdieu, P. (1980). Le capital social: Notes provisoires. *Actes de la Recherche en Sciences Sociales*, 3, 2–3.

Bourdieu, P. (1983/1986). The forms of capital. In J. G. Richardson (Ed.), *Handbook of theory and research for the sociology of education* (pp. 241–258). Westport, CT: Greenwood Press.

Canzler, W., & Knie, A. (2016). Mobility in the age of digital modernity: Why the private car is losing its significance, intermodal transport is winning and why digitalisation is the key. *Applied Mobilities*, 1(1), 56–67. https://doi.org/10.1080/23800127.2016.1147781

Carrasco, J. A., & Cid-Aguayo, B. (2012). Network capital, social networks, and travel: An empirical illustration from Concepción, Chile. *Environment and Planning A*, 44, 1066–1084.

Carrasco, J. A., Hogan, B., Wellman, B., & Miller, E. J. (2008). Collecting social network data to study social activity-travel behavior: An egocentric approach. *Environment and Planning B: Planning and Design*, 35(6), 961–980.

Castells, M. (2001). *The internet galaxy: Reflections on the internet, business and society*. Oxford: Oxford University Press.

Cidell, J. (2017). Aero-automobility: Getting there by ground and by air. *Mobilities*, 12(5), 692–705. https://doi.org/10.1080/17450101.2016.1240318

Coleman, J. S. (1990). *Foundations of social theory*. Cambridge, MA: Harvard University Press.

Couldry, N., & Hepp, A. (2017). *The mediated construction of reality*. Cambridge, UK; Malden, MA: Polity Press.

Cwerner, S., Kesselring, S., & Urry, J. (Eds.). (2009). *Aeromobilities: International library of sociology*. London; New York, NY: Routledge.

Dennis, K. L. (2013). Mobility futures: Moving on and breaking through on an empty tank. In S. Witzgall, G. Vogl, & S. Kesselring (Eds.), *New mobilities regimes: The analytical power of the social sciences and arts* (pp. 331–354). Burlington, VT: Ashgate Publishing Ltd.

Dennis, K., & Urry, J. (2009). *After the car*. Cambridge: Polity Press.

Diez, W. (2018). *Wohin steuert die deutsche Automobilindustrie?* (2., überarbeitete und aktualisierte Auflage). Berlin; Boston, MA: De Gruyter Oldenbourg. Retrieved from www.degruyter.com/search?f_0=isbnissn&q_0=9783110481150&searchTitles=true

Dodge, M., & Kitchin, R. (2004). Flying through code/space: The real virtuality of air travel. *Environment and Planning A, 36*(2), 195–211.

Elliott, A. (2016). *Identity troubles: An introduction.* London; New York, NY: Routledge Taylor & Francis Group.

Elliott, A. (2018). *The culture of AI: Everyday life and the digital revolution.* Milton: Routledge.

Elliott, A., & Urry, J. (2010). *Mobile lives: Self, excess and nature: International library of sociology.* London; New York, NY: Routledge.

Fischer, C. S. (1982). What do we mean by 'Friend'? An inductive study. *Social Networks, 3,* 287–306.

Frändberg, L., & Vilhelmson, B. (2003). Personal mobility: A corporeal dimension of transnationalisation: The case of long-distance travel from Sweden. *Environment and Planning A, 35,* 1751–1768.

Frei, A., & Axhausen, K. W. (2007). Size and structure of social network geographies. *Arbeitsberichte Verkehrs und Raumplanung, 439,* IVT, ETH Zürich, Zürich,

Freudendal-Pedersen, M., & Kesselring, S. (2018a). Networked urban mobilities. In M. Freudendal-Pedersen & S. Kesselring (Eds.), *Networked urban mobilities series: Vol. 1: Exploring networked urban mobilities: Theories, concepts, ideas* (1st ed., pp. 1–18). New York, NY: Routledge.

Freudendal-Pedersen, M., & Kesselring, S. (2018b). Sharing mobilities: Some propaedeutic considerations. *Applied Mobilities, 3*(1), 1–7. https://doi.org/10.1080/23800127.2018.1438235

Freudendal-Pedersen, M., Kesselring, S., & Servou, E. (2019). What is smart for the future city? Mobilities and automation. *Sustainability, 11*(1), 1–25.

Geels, F. W. (Ed.). (2012). *Automobility in transition? A socio-technical analysis of sustainable transport: Routledge studies in sustainability transitions* (Vol. 2). New York, NY: Routledge.

Geertz, C. (1973). *The interpretation of cultures.* New York, NY: Basic Books.

Giddens, A. (1984). *The constitution of society: Outline of the theory of structure.* Berkeley, CA: University of California Press.

Gustavson, P. (2012). Travel time and working time: What business travellers do when they travel, and why. *Time & Society, 12*(2), 203–222.

Hajer, M. A., & Dassen, T. (2014). *Smart about cities: Visualising the challenge for 21st century urbanism.* Rotterdam: Nai010 Publishers. Retrieved from www.nai010.com/en/component/zoo/item/smart-about-cities

Hilgenberg, J. (2017). Die Lehren aus Dieselgate: Anhaltende Überschreitung der Stickstoffoxid-Grenzwerte. In *Politische Ökologie: Vol. 148: Zukunftsfähiges Deutschland: Wann, wenn nicht jetzt?* (pp. 140–143). München: oekom verlag.

Huszti, E., Dávid, B., & Vajda, K. (2013). Strong tie, weak tie and in-betweens: A continuous measure of tie strength based on contact diary datasets, 9th Conference on Applications of Social Network Analysis (ASNA). *Procedia – Social and Behavioral Sciences, 79,* 38–61.

Kesselring, S. (2006). Pioneering mobilities: New patterns of movement and motility in a mobile world. *Environment and Planning A, 38*(2), 269–279.

Kim, J., Rasouli, S., & Timmermans, H. J. P. (2018). Social networks, social influence and activity-travel behaviour: A review of models and empirical evidence. *Transport Reviews, 38*(4), 499–523.

Kowald, M., & Axhausen, K. W. (2010, September). The structure and spatial spread of ego-centric leisure networks: Conference paper for Applications of Social Network Analysis (ASNA), Zurich, September 15–17, 2010. In *7th International Conference on Applications of Social Network Analysis, ASNA 2010, Abstracts* (Vol. 651, pp. 65–66). ETH.

Kowald, M., & Axhausen, K. W. (2016). *Social networks and travel behaviour*. London: Routledge.

Kowald, M., Frei, A., Hackney, J. K., Illenberger, J., & Axhausen, K. W. (2009, September). Collecting data on leisure travel: The link between leisure acquaintances and social interactions. Conference paper (STRC), 9th Swiss Transport Research Conference, Ascona, Switzerland: Monte Verità, September 9–11.

Kowald, M., van den Berg, P., Frei, A., Carrasco, J. A., Arentze, T., Axhausen, K. . . . Wellman, B. (2013). Distance patterns of personal networks in four countries: A comparative study. *Journal of Transport Geography, 31,* 236–248.

Larsen, J., Urry, J., & Axhausen, K. (2006). Geographies of social networks: Meetings, travel and communications. *Mobilities, 1*(2), 261–285.

Lash, S. (1999). *Another modernity: A different rationality*. Oxford: Blackwell Publishers.

Lin, N. (2001). *Social capital: A theory of structure and action*. London; New York, NY: Cambridge University Press.

Lübbe, H. (1993). Mobilität: Verkürzter Aufenthalt in der Gegenwart. In H. Schaufler (Ed.), *Mobilität und Gesellschaft* (pp. 141–153). München: mvg.

Lyons, G., Jain, J., & Holley, D. (2007). The use of travel time by rail passengers in Great Britain. *Transportation Research Part A, 41*(1).

Lyotard, J. (1984). *The postmodern condition: A report on knowledge*. Manchester: Manchester University Press.

McCallister, L., & Fischer, C. S. (1978). An approach for surveying personal networks. *Sociological Methods & Research, 7,* 131–148.

Mom, G. (2015). *Atlantic automobilism: Emergence and persistence of the car, 1895–1940: Explorations in mobility* (Vol. 1). New York, NY: Berghahn Books.

Morley, D. (2017). *Communications and mobility: The migrant the mobile phone and the container box*. Hoboken, NJ; Chichester, West Sussex: Wiley Blackwell.

Ohnmacht, T. (2006). Mapping social networks in time and space. *Arbeitsberichte Verkehr und Raumplanung, 341,* IVT, ETH, Zürich.

Putnam, R. D. (2000). *Bowling alone: The collapse and revival of American community*. New York, NY: Simon & Schuster.

Rainie, L., & Wellman, B. (2012). *Networked: The new social operating system*. Cambridge, MA: Massachusetts Institute of Technology Press.

Sachs, W. (1992). *For love of the automobile: Looking back into the history of our desires*. Berkeley, CA: University of California Press.

Schönfelder, S., & Axhausen, K. W. (2003). Activity spaces: Measures of social exclusion? *Transport Policy, 10,* 273–286.

Schutz, A. (1962). *The problem of social reality*. The Hague: Nijhoff.

Schutz, A. (1971). *Collected papers* (Vols. 1 and 2). The Hague: Nijhoff.

Sharmeen, F. (2015). Dynamics of social networks and activity travel behaviour. In *Department of the Built Environment*. Eindhoven: Eindhoven University of Technology. Retrieved from https://pure.tue.nl/ws/files/3964222/780949.pdf

Silverstone, R. (Ed.). (2005). *Media technology and everyday life in Europe*. Aldershot: Ashgate Publishing Ltd.

Silvis, J., Niemeier, D., & D'Souza, R. (2006). Social networks and travel behavior. Report from an Integrated Travel Diary, 11th International Conference on Travel Behavior Research, Kyoto.

Simmel, G. (Ed.). (1923). *Soziologie: Untersuchungen über die Formen der Vergesells-chaftung*. München; Leipzig: Duncker & Humblot.

Simmel, G. (1964). *The sociology of Georg Simmel*. New York, NY: The Free Press.

Simmel, G. (1971). *On individuality and social forms: Selected writings: Heritage of sociology*. Chicago, IL: University of Chicago Press.

Spinney, J., & Lin, W.-I. (2018). Are you being shared? Mobility, data and social relations in Shanghai's Public Bike Sharing 2.0 sector. *Applied Mobilities, 3*(1), 66–83. https://doi.org/10.1080/23800127.2018.1437656

Thrift, N. (2004). Movement-space: The changing domain of thinking resulting from the development of new kinds of spatial awareness. *Economy & Society, 33*(4), 582–604.

Tönnies, F. (1957). *Community and society*. East Lansing, MI: Michigan State University Press.

Turkle, S. (2011). *Alone together: Why we expect more from technology and less from each other*. New York, NY: Basic Books. (Paperback first published).

Urry, J. (2000). *Sociology beyond societies: Mobilities for the twenty-first century*. London: Routledge.

Urry, J. (2002). Mobility and proximity. *Sociology, 36*(2), 255–274.

Urry, J. (2003). Social networks, travel and talk. *British Journal of Sociology, 54*(2), 155–175.

Urry, J. (2004). The 'system' of automobility. *Theory, Culture & Society, 21*(4–5), 25–39.

Van den Berg, P., Arentze, T. A., & Timmermans, H. (2009). Size and composition of ego-centered social networks and their effect on geographic distance and contact frequency. *Transportation Research Record: Journal of the Transportation Research Board, 2135*, 1–9. Washington, DC.

Van den Berg, P., Arentze, T. A., & Timmermans, H. (2010, July). A path analysis of social networks, ICT use and social activity-travel patterns in the Netherlands. Conference Paper, 12th WCTR, July 11–15, 2010, Lisbon, Portugal.

Vilhelmson, B., & Thulin, E. (2008). Virtual mobility, time use and the place of the home. *Tijdschrift voor Economische en Sociale Geografie, 99*, 602–618.

Wasserman, S., & Faust, K. (1994). *Social network analysis, methods and applications*. New York, NY: Cambridge University Press.

Weber, M. (1922). *Economy and society: An outline of interpretive sociology*. New York, NY: Bedminster Press.

Wellman, B. (1979). The community question: The intimate networks of East Yorkers. *American Journal of Sociology, 84*, 1201–1231.

Wellman, B. (2001, October). Little boxes, glocalization, and networked individualism. In *Kyoto workshop on digital cities* (pp. 10–25). Berlin; Heidelberg: Springer.

Wellman, B., & Haythornthwaite, C. (Eds.). (2002). *The internet in everyday life*. Malden, MA: Blackwell Publishers.

Wellman, B., & Marin, A. (2011). Social network analysis: An introduction. In P. Carrington & J. Scott (Eds.), *Handbook of social network analysis* (pp. 11–25). Thousand Oaks, CA: Sage Publications.

Wessels, B. (2010). *Understanding the internet: A socio-cultural perspective*. Basingstoke: Palgrave Macmillan.

Wessels, B. (2014). *Exploring social change: Process and context*. Basingstoke: Palgrave Macmillan.

Winnicott, D. W. (1997). *Playing and reality*. London: Routledge. (Reprint).

3 Opinion dynamics and complex networks

*Silvana Stefani, Maria Candelaria Gil-Fariña
and Marija Mitrović Dankulov*

Introduction

Where do beliefs and opinions come from? During our lifetime and based on our own learning experience, we acquire beliefs and opinions. Much of this learning takes place within families, where children learn certain basic principles and beliefs (see e.g. Boyd & Richerson, 2005). Later on, learning becomes "social", through a process of learning whereby individuals obtain information and update their beliefs and judgements as a result of their own experiences, their observations of others' actions and experiences, education, the communication with others, news from media sources, from political leaders and celebrities. As a consequence, while the process of learning by an individual from his or her experience can be viewed as an "individual" learning process, it also has an explicitly "social" character.

Opinion formation by social influence is the process by which individuals adapt their opinion, revise their beliefs, or change their behaviour as a result of social interactions with other people. In our strongly interconnected society, social influence plays a prominent role in many self-organized phenomena such as the spread of ideas and innovations, such as electric cars, or mobility choices. What it becomes relevant is the process under which opinion formation leads to consensus (a general agreement), polarization (two or more different opinions emerge) or fragmentation (opinions are scattered). This process is called opinion dynamics.

Opinion formation among individuals and the resulting dynamics that is induced in a social network has been extensively studied, both from the perspective of analytic modelling and experimental psychology (Gerard & Orive, 1987; Boyd & Richerson, 2005; Acemoglu, Ozdaglar, & Parandeh Gheibi, 2009; Das, Gollapudi, & Munagala, 2014; Galam, 2011; Raafat, Chater, & Frith, 2009; Yildiz, Acemoglu, Ozdaglar, Saberi, & Scaglione, 2011; Jalili, 2013; Yaniv, 2004). Examples from biology and animal behaviour show similarities with humans, like the herding effect caused by limited attention (Raafat et al., 2009; Clark & Dukas, 2003; Dukas, 2004; Kameda, Inukai, Wisdom, & Toyokawa, 2014; Cont & Bouchaud, 2000). Herding refers to an alignment of thoughts or behaviours of individuals in a group. Most important, such convergence often emerges through local interactions among agents rather than some purposeful coordination by a central authority or a leading figure in the group. In other words, the apparent coordination of the herd is an emergent pattern of local interactions

It is known that several complementary influence processes shape opinion formation in society. One of these processes is called informational influence, where users in the absence of necessary information, or on the contrary exposed to much information, seek the opinions of their neighbours (the general term for indicating individuals directly connected to user) in order to update their beliefs (Hegselmann & Krause, 2002; Das et al., 2014, Dong et al., 2017).

Informational influence may be used by opinion leaders to drive consensus or to polarize opinions.

In 2005, an important project started for a 57-km high-velocity train tunnel connecting Torino (Italy) and Lyon (France), through Val di Susa. The project is part of Corridor 3, Mediterraneum, in the European Railway System. However, in Italy a political campaign against it rapidly grew, led by some important (minority at that time) leaders, such as 5-Stars Grillo, a leader much followed and very smart in communication. Information about the tunnel was very poor and never given at technical level; naturally, people relied on opinion leaders such as Grillo. There were riots and manifestations. As a result, the work for the tunnel stopped for 10 years and started back in 2016; after that, the same parties (now a majority), who were once against, decided to be pro-tunnel.

Under the informational influence, individuals update their opinion based on the information learned from neighbours (as are called the individuals that are directly connected). Finding themselves in the same situation under uncertainty about the problem at stake, they will be receptive to one another's ideas. Experiments show that if two people communicate (and do not strongly dislike mutual opinions), their opinion will change linearly with the difference between them. In practice, the final opinion will be the average of the two, and eventually, they will share the same opinion. The same principles are valid for a group or even a population. The shape of the connections between individuals gives the topology of the network (Bruggemann, 2018; Friedkin & Johnsen, 2011). The time evolution of opinions in the network can be modelled in terms of convergence time, and an eventual pattern will finally emerge.

In the dynamics of opinion formation, consensus may converge toward choices that may not necessarily be optimal for the society involved, like rejection or delay in the acceptance of new technologies, without apparent technical reasons.

The Theory of the Wisdom of the Crowd, or Vox Populi, dates back to Galton (1907) and discusses the trustworthiness of popular judgement[1] (Surowiecki, 2005). On the other hand, the 1928 Thomas theorem "If men define situations as real, they are real in their consequences" may be quoted in situations of opinion dynamics, where people recognize as true some news that, later on, are recognized as false. What follows is a motivating example, where Thomas theorem and Merton conjecture find their application (Merton, 1995).

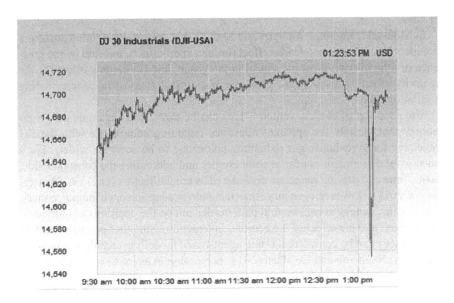

Figure 3.1 23 April DJ30 Industrial Index freefall – 1-minute intraday

On 23 April 2013, at 1:07 p.m., the U.S. Stock Market dropped by about 1% in approximately two minutes, recovering soon afterwards in about the same time. Figure 3.1 shows a snapshot of the DJ 30 Industrial Index one-minute intraday time series during those minutes. The cause of the sudden drop was false news: at 1:07 p.m., the Associated Press tweeted that President Obama had been reported injured in two explosions at the White House. The news was false, probably due to hacker access to the Associated Press account. The tweet was immediately shared many times (about 4,000 in less than one minute and more in the next minutes). The "flash-crash" lasted two minutes and led to a 1% drop in the index, that is a loss of 130 billion dollars. What caused the drop was probably due to the reliability of the Associated Press agency, recognized as an authority that led investors to immediately trust the consequences of such a tweet, without verifying its truthfulness. The example shows how the system is vulnerable and open to opinion shifts, regardless of the truthfulness of the underlying event.[2]

Whether accurate or not, individual views, opinions and beliefs have essential effects on the aggregate behaviour of the whole system.

Summing up, opinion formation is not always the result of optimal information processing. This attitude is confirmed by the famous sentence "A wealth of information creates a poverty of attention" by Simon (1971). When we are faced with an important decision to take, most times, we find it impossible to retrieve all information we need, and we refer to someone else, neighbours, friends, experts, opinion leaders and even celebrities to get clues or advice.

"Referring to" means establishing connections and an information network creates.

In Moussaïd, Kammer, Analytis and Neth (2013), two major opinion attractors are identified: (1) opinion leader effect (or the expert effect), induced by the presence of a highly confident individual in the group, and (2) herding effect (or the majority effect), caused by the presence of a critical mass of laypeople sharing similar opinions.

The adoption of electric vehicles (EV) can be taken as an example of a technology that, despite the apparent relevance regarding compliance with the EU objective for greenhouse gas reduction, takes time to be accepted widely. For a successful dissemination of new technologies and addressing the consumer side, costs, time and driving range are considered as crucial factors (Ozaki & Sevastyanova, 2011). However, even more than technology adoption or economic matters, large-scale changes in behavioural patterns depend on the decisions of individual consumers, and the effect of the complex interaction with other people regarding experience must be considered. Using agent-based modelling techniques, in Wolf, Schröder, Neumann and De Haan (2014) persuasion in agent-to-agent communication is proved to be a critical factor in the dissemination of EV cars. Moreover, consumer acceptance of electric cars is an example of the importance of social influence as a trigger of behavioural change (Rasouli & Timmermans, 2016). Furthermore, while correct parameters are strategic to provide policy-makers with an accurate decision-making tool, little has been done to explore the dynamics of the process by which the market share in travelling modes adjusts to changes in the environment stemming from such policy decisions (e.g. decrease in travel time, increases in parking costs). It is essential to evaluate the process of these market shifts since it will take time for the entire population to become aware of changes in the attributes. Instead, individuals may learn about these changes through various social and interpersonal mechanisms over time, such as word of mouth, the mass media, direct experience and beliefs (Kozuki, 2007; Timmermans et al., 2014).

In an important project that is undertaken at national or even local level, such as building a new road or a new, fast railway line, the public manager must keep an eye on how the information network is structured and start an active campaign recruiting opinion leaders, experts and celebrities, who are proved to play a strategic role in driving towards a decision or another.

In the network, the importance of one agent can be modelled by a weight which the other agents put on the opinions of him. The more important is the agent (for instance an opinion leader), the higher is the weight the others put on him, while on the contrary the opinion leader, especially if self-confident, does not put much weight to the other's opinion. Each agent places weight to colleagues, friends, family members, experts and any other agent considered the worth of attention. A two-entry table, called the attention matrix, collects all weights. The process of exchanging opinions goes on with time, and a general opinion eventually takes shape, obtained by averaging, again and again, all opinions together. A dynamical process in discrete time generates. Intuitively, one may expect that this process of repeatedly averaging opinions will bring newly formed opinions of different agents closer to each other until they flow into a final configuration of opinions (DeGroot, 1974; Hegselmann & Krause, 2002).

How to model opinion dynamics

Assume that a survey has been conducted in the metropolitan area of Bologna for the realization of a new tramway line. In principle, people are not against it, but the tramway will pass through busy neighbourhoods, cutting parking places and obliging residents to find parking far away. Furthermore, many roads will be closed or reduced in width and driving will be difficult. Parking places will disappear, and commuters driving to the city centre will be obliged to park at the periphery and take the tramway. The survey is composed of just one question: "Are you pro (Y), against (N), indifferent (I) to the new tramway line?" After the survey, the debate starts. Results are not clear-cut. There is a slight preference for Y, but I is very high, and the undecided may eventually turn one way or the other. In Bologna, (approximately 55,000 inhabitants in the city centre), a well-known public opinionist lives in the city centre and is strongly pro-tramway. The survey is expensive and repeating it, again and again, makes no sense, so, taking the opinion dynamics approach and following DeGroot (1974), the municipality, based on demographic data and analysis of social media, can identify six homogeneous groups of citizens and build an attention matrix. The entries (i,j) in this attention matrix are the weights that the i-th group assigns as importance to the j-th group. The groups are (1) opinion leader and his followers, (2) followers of the opinion leader but with own ideas, (3) indexers (so called as they pay the same attention to every group in the area), (4) sympathizers for the opinion leader but with own ideas and slightly interested to some of the other groups, (5) those interested in the opinion leader but more interested to Group 2 and (6) not at all interested in the opinion leader but attentive to Group 5. The sum by row gives the total attention of each group, and this sum is 1 (100% of personal attention). For example, Group 1 is almost entirely self-centred (or stubborn; 90%) and pays a small amount of attention only to group 2 (10%); group 6 gives 50% of its attention to group 5 and 50% to itself (see Table 3.1).

Table 3.1 Attention matrix

Group	1	2	3	4	5	6
1	0.9	0.1	0	0	0	0
2	0.88	0.12	0	0	0	0
3	0.167	0.167	0.167	0.167	0.167	0.167
4	0.55	0.33	0.1	0.01	0.01	0
5	0.25	0.75	0	0	0	0
6	0	0	0	0	0.5	0.5

After the survey, answers are collected for each group, and the probability distributions for each group are computed (Table 3.2). For example, in Group 3, composed of 15000 respondents, 25% were pro tramway, 75% indifferent and none against.

The only one who has practically no doubts is Group 1, the opinionist and his followers: they are 99% pro tramway. However, of the totality of respondents, more than 65% are indifferent or against the tramway, and this proportion is too high for the municipality.

The problem is to reach a consensus, or at least a vast majority pro-tramway, and better have it fast for making the works start on time. After 11 units of time (it can be months), without any unexpected event that may change the attention matrix, the distribution becomes stable (Table 3.3).

Mathematical details are in Appendix 1.

Results are quite clear. After discussions, public meetings and debates, the opinionist, even though belonging to a small group in numbers (less than 10% of the total respondents), drives the opinions. The connection between groups allows the opinion leader effect to indirectly touching also those who were not interested in his opinions. The last group, not at all interested in the opinionist, follows the majority, being affected by Group 5.

The municipality may start works of the tramway line, with the confidence of being well supported.

Table 3.2 Probability distribution for each group

Group	Y	I	N	# respondents
1	0.99	0	0.01	4,000
2	0.33	0.33	0.33	15,000
3	0.25	0.75	0	15,000
4	0.25	0.75	0	6,000
5	0.6	0.2	0.2	4,500
6	0	0.3	0.7	10,000
				55,000

Table 3.3 Stable probability distribution

Group	Y	I	N
1	0.9230	0.0340	0.0430
2	0.9230	0.0340	0.0430
3	0.9224	0.0343	0.0433
4	0.9229	0.0341	0.0431
5	0.9230	0.0340	0.0430
6	0.9220	0.0344	0.0436

It is easy to see that in the same situation but in the absence of an opinionist, the distribution eventually becomes uniform; that is opinions converge to an equal share for Y, I, N, with no help for the municipality.

This simple example shows how opinion dynamics work. In the case of some unexpected events or news that may change the mutual attention of groups, the attention matrix changes accordingly, the procedure is repeated, and a new equilibrium is possibly reached.

The topology matters and affects the velocity of convergence that is the number of time units for the distribution to become stable. Small and almost isolated communities slow convergence; self-centred (stubborn) individuals slow convergence but may contribute to a consensus. Stubborn opinionists (inflexible) of opposite views may delay convergence indefinitely (see also Galam, 2011; Yildiz et al., 2011; Acemoglu et al., 2009).

In Golub and Jackson (2010), learning and influence are studied in a setting where agents receive noisy independent signals about the real value of a parameter (the probability distribution) and then communicate in a social network. The agents update their beliefs over time by repeatedly taking weighted averages of their neighbours' opinions. Conditions are identified, determining whether the beliefs of all agents in large societies converge to the truth value of the variable, the truth value is defined as a value θ, the wisdom of the crowd. The authors show that such convergence to truth is obtained if and only if the influence of the most influential agent in the society vanishes as the society grows. Obstructions can prevent such convergence, including the existence of prominent groups which receive a disproportionate share of attention.

In Hegselmann and Krause (2002) a time-dependent version and a non-linear version with the bounded confidence of the agents are introduced. Bounded confidence arises when an agent i takes into account only those neighbours/agents j whose opinions differ from his own not more than a certain confidence level. If the neighbour's opinion differs "too much" (i.e. beyond the confidence level) from the agent's opinion, the agent shifts his/her attention to others and the attention matrix changes. Thus, bounded confidence is modelled by an attention matrix that is varying with time k and that does not have constant entries. The bounded confidence model is non-linear and may even generate chaotic trajectories.

In Rogers and Murillo (2009), in a bounded confidence model, control of opinions is studied. The main conclusions are that averaging models capture realistic opinion dynamics, strategies exist to control opinions, but convergence is slow.

The tools

The investigation of social networks has regularly attracted contributions from applied mathematicians and social scientists over the last decades. Graph theory, matrix algebra and computer science have applications to such investigations (see Chapter 2, "Social Networks Theory: Definitions and Practice", in this same book). Much ongoing interest is focusing on dynamic models of structural change and a broad range of dynamic processes unfolding over static networks; examples include the study of social learning, opinion formation and information propagation. The study of dynamic models directly addresses one of the key problems of the field, which is to understand the social implications in a society based on relevant dynamical states of the network.

In Muoneke (1987), a necessary condition is found for a non-negative reducible matrix to have a stochastic power, that is to find an integer k such that A^k is stochastic. Various aspects of the theory of random walks on graphs are relevant here and are surveyed in Lovász (1993) and in Meyn and Tweedie (2005). In particular, estimates on the critical parameters of access time, commute time, cover time and mixing time are discussed. Algorithmic applications of random walks are introduced, in particular to the problem of sampling.

Classical results on matrix algebra, among which positive and non-negative matrices, primitive and stochastic matrices and the Perron Frobenius Theorem are illustrated in Meyer (2000). The DeGroot model is extended and clarified in Berger (1981). DeGroot studies how to reach a consensus in a dynamical model about the value of a parameter θ of which each has a probability estimating distribution. According to DeGroot, the consensus is reached if and only if all individuals eventually have the same perception (same probability distribution) for the parameter θ. In Lorenz (2007), the opinion dynamics case with bounded confidence, changing confidence and with matrices with positive diagonal is discussed. Simulations are provided.

Opinion dynamics: fields of application

The purpose of this section is to show how opinion dynamics is used and applied in different fields. It provides useful insights for extending the research on opinion dynamics to fields like transportation where contributions on the specific topic are still limited.

Social sciences and elections

The most relevant application of opinion dynamics techniques is in electoral situations or public debates and, generally speaking, in situations in which unanimity or a significant majority is required. In Yildiz et al. (2011), opinion dynamics is studied in a social network with stubborn agents. They generalize the classical voter model by introducing stubborn agents who are self-centred and do not pay much attention to others. It is shown that the presence of stubborn agents with opposing opinions precludes convergence to consensus; instead, opinions converge in distribution with disagreement and fluctuations. They also study the problem of "optimal placement of stubborn agents", where the location of a fixed number of stubborn agents is chosen to have the maximum impact on the long-run, expected opinions of agents.

A simple stochastic framework of a collective process of opinion formation by a group of agents who face a binary decision problem is studied in Lux (2009). This approach can be used to estimate the parameters of the opinion formation process from aggregate data. The combined effects of collective beliefs and individual inflexibility in the dynamics of public debate are investigated in Galam (2011). Galam's results hint on how to design possible strategies to win a public debate. In particular, it has been shown how the existence of collective beliefs is a major factor in the direction taken by a public debate. To counterbalance this strong bias, the making of stubborn (the "inflexibles") agents is efficient but requires a substantial fraction of them. Since supposedly and at least in the short run the collective beliefs are not given to modifications, the best approach for one opinion to win is to focus on getting as many as possible inflexibles along its side.

Economics and finance

The tendency of market participants to imitate each other is modelled in Cont and Bouchaud (2000). Copycat investing, as the name implies, refers to the strategy of replicating the investment decisions of famous gurus or investment managers. Whether copycat investing is a viable investment strategy, it is still to be demonstrated since the evidence is somewhat mixed. Kim (2005) describes the behaviour of copycats of the famous guru Warren Buffett. In an opinion-dynamics setting, copycats replicate the behaviour of their guru that is they pay attention to the guru only and not even to themselves.

Transportation

Contributions of opinion dynamics and consensus are rare in this field of application.

In the market adoption of a new rail-freight service, in Kozuki (2007) a model is discussed that utilizes social and learning mechanisms to, first, explore the underlying dynamics of opinion formation and propagation and then applies those mechanisms to an application of freight-mode choice to investigate the effect that opinions have on modal choice. Investigating opinion dynamics in the context of a choice framework may also offer insight into how information about the attributes, and, more important, the changes in the attributes, spreads amongst a population.

A favourable opinion about a mode does not necessarily result in choosing that mode; instead, a favourable opinion towards an alternative may encourage an individual to try the alternative only for a trial period. Once the trial period is complete (or, in other words, if the initial conditions of the systems change or the attention shifts to another media or information device), the individual may decide to either continue utilizing the alternative or go back to the previous alternative (Rogers, 2003; Garling, 1998).

The development of intelligent transportation systems (ITS), traffic management centres and real-time traffic navigation have made real-time information on transportation facilities available to users. Information provided by these technological advances will redirect users to the available facilities (in the case of congestion) or alternative services. Again, it may not be easy to determine whether the information provided is actually affecting an individual's opinion and ultimately his choice as a stable alternative (Wu & Huberman, 2005). For example, commuters share information about accidents, failures, delays or "everything is all right". Understanding how users in a network update their opinions based on their neighbours' opinions, as well as what global opinion structure is implied when users iteratively update opinions, is essential in the context of viral marketing and information dissemination, as well as targeting messages to users in the network. While travelling back and forth, commuters get to know each other, often not personally, but soon or later some central actors emerge that are considered more reliable than others, thereby contributing more than others to the information spreading process. After reading the neighbours' opinions, and among them the leaders' opinions, commuters will stick with their own modal choice or change their itinerary or change mode.

In Timmermans et al. (2014), an extensive review of dynamics in travel behaviour is presented and discussed. Han, Arentze, and Timmermans (2010) propose a dynamic model based on principles of social learning and knowledge transfer in modelling the impact of social networks. Using agent-based modelling, Hačkney and Marchal (2009) develop a multi-agent representation, including the dynamics of social networks, by adding and deleting links and assuming a maximum number of contacts for each agent. Arentze and Timmermans (2008) discuss a theoretical framework to model social networks, social interactions and travel behaviour. Most of this work is based on numerical simulations. Sharmeen, Arentze, and Timmermans (2010, 2014a, 2014b) investigate the evolution of social networks in correspondence of life cycle. They argue that social networks change with life-cycle events and affect activities and travel needs.

In public transport management and planning, including active participation of citizens and stakeholders from the beginning of transport decision-making processes is widely recognized as a precondition to avoid the failure of projects/policies/plans as a consequence of a lack of consensus. Appropriate methods and tools are needed to support participation processes towards well-thought-out and shared solutions. Transport systems require special attention since their planning affects the livability and economy of a community. Thanks to the new technologies, it is easy to exchange and express opinions and interact via the web (e.g. via blog, forum, social media). This kind of online participation is consistent with the principles of the "smart cities", where information flows fast and efficiently and creates networks that are highly interconnected. However, since this kind of participation is generally self-organized and not monitored, it can hide some pitfalls (Le Pira, Inturri, Ignaccolo, Pluchino, & Rapisarda, 2015).

It is clear that the impact of information is relevant, and the correct use of it by a local decision maker may have a profound effect on mobility and travelling flows. It is also important to consider whether the information diffusion is ultimately affecting an individual's choice, especially when considering market adoption of a new service and to what extent attribute distortion plays a role in choice mechanisms (Kozuki, 2007). Exploring these mechanisms of information exchange offers insight to how individuals may react to changes in the level of service of a particular alternative and, more important, how over time the information about a new alternative or service may propagate through a population and whether and how the general opinion will converge to a stable pattern. Opinion-dynamics techniques may be a valuable tool for addressing those critical issues.

Notes

1 An example, taken from Galton (1907), is about a weight judging competition that was carried out at a fair exhibition. A fat ox had been selected; competitors bought stamped cards and gave secretly their own estimates of the ox's weight. The one who gave the closest estimate to the real weight would be the winner. It happened that the median of the estimates was within 1% of the real value, but estimates differed quite substantially. Galton stated that according to the democratic principle of "one vote, one man", the middlemost estimate expresses the *vox populi*.
2 H. Moore and Dan Robertson, "AP Twitter hack causes panic on Wall Street and sends Dow plunging", www.theguardian.com/business/2013/apr/23/ap-tweet-hack-wall-street-freefall.

References

Acemoglu, D., Ozdaglar, A., & Parandeh Gheibi, A. (2010). Spread of (mis)information in social networks. *Games and Economic Behavior, 70*(2), 194–227.

Arentze, T. A., & Timmermans, H. J. P. (2008). Social networks, social interactions and activity-travel behavior: A framework for microsimulation. *Environment and Planning B, 35*(6), 1012–1027.

Berger, R. L. (1981). A necessary and sufficient condition for reaching a consensus using DeGroot's method. *Journal of the American Statistical Association, 76*(374), 415–418. doi:10.1080/01621459.1981.10477662

Boyd, R., & Richerson, P. J. (2005). *The origin and evolution of cultures.* Oxford: Oxford University Press. ISBN-10 019518145X, ISBN-13 9780195181456

Bruggemann, J. (2018). Consensus, cohesion and connectivity. *Social Networks, 52,* 115–119.

Clark, C. W., & Dukas, R. (2003). The behavioral ecology of a cognitive constraint: Limited attention. *Behavioral Ecology, 14*(2), 151–156.

Cont, R., & Bouchaud, J. P. (2000). Herd behavior and aggregate fluctuations in financial markets. *Macroeconomic Dynamics*, *4*(2), 170–196.

Das, A., Gollapudi, S., & Munagala, K. (2014). Modeling opinion dynamics in social networks. WSDM '14 *Proceedings of the 7th ACM International Conference on Web Search and Data Mining*, 403–412. doi:10.1145/2556195.2559896

DeGroot, M. H. (1974). Reaching a consensus. *Journal of the American Statistical Association*, *69*(345), 118–121.

Dong, Y., Ding, Z., Martínez, L., & Herrera, F. (2017). Managing consensus based on leadership in opinion dynamics. *Information Sciences*, *397–398*, 187–205.

Dukas, R. (2004). Causes and consequences of limited attention. *Brain Behavior and Evolution*, *63*, 197–210.

Friedkin, N., & Johnsen, E. (2011). *Social influence network theory: A sociological examination of small group dynamics*. Cambridge: Cambridge University Press.

Galam, S. (2011). Collective beliefs versus individual inflexibility: The unavoidable biases of a public debate. *Physica A*, *390*(17), 3036–3054.

Galton, F. (1907). Vox populi. *Nature*, *75*, 450–451.

Garling, T. (1998). Reintroducing attitude theory in travel behavior research: The validity of an interactive interview procedure to predict car use. *Transportation*, *25*(2), 129–146.

Gerard, H. B., & Orive, R. (1987). The dynamics of opinion formation. *Advances in Experimental Social Psychology*, *20*, 171–202.

Golub, B., & Jackson, M. O. (2010). Naïve learning in social networks and the wisdom of crowds. *American Economic Journal: Microeconomics*, *2*(1), 112–149.

Hackney, J. K., & Marchal, F. (2009, January). Model for coupling multi-agent social interactions and traffic simulation. Compendium of papers of the 88th Annual Meeting of the Transportation Research Board (pp. 11–15). Washington, DC.

Han, Q., Arentze, T. A., & Timmermans, H. J. P. (2010, July). Habit formation and effective responses in location choice dynamics. *Proceedings of the 12th World Conference on Transportation Research*, July 11–15, Lisbon.

Hegselmann, R., & Krause, U. (2002). Opinion dynamics and bounded confidence: Models, analysis and simulation. *Journal of Artificial Societies and Social Simulation*, *5*(3).

Jalili, M. (2013). Social power and opinion formation in complex networks, *Physica A*, *392*(4), 959–966.

Kameda, T., Inukai, K., Wisdom, T., & Toyokawa, W. (2014). *The concept of herd behaviour: Its psychological and neural underpinnings*. Retrieved from www.tatsuyakameda.com/_src/614/chapter-2.pdf

Kim, J. J. (2005, April). Wisdom fund' aims to imitate Berkshire's Holdings. *The Wall Street Journal*, April 28. Retrieved from www.wsj.com/articles/SB111465010297619054

Kozuki, A. (2007). *Modeling the dynamics of opinion formation and propagation: An application to market adoption of transportation services*. DRUM, Digital Repository University of Maryland, http://hdl.handle.net/1903/7588

Le Pira, M., Inturri, G., Ignaccolo, M., Pluchino, A., & Rapisarda, A. (2015). Simulating opinion dynamics on stakeholders' networks through agent-based modeling for collective transport decisions. *Procedia Computer Science*, *52*, 884–889.

Lorenz, J. (2007). *Convergence of products of stochastic matrices with positive diagonals and the opinion dynamics background*. doi:10.1007/11757344

Lovász, L. (1993). Random walks on graphs: A survey. In *Bolyai Society Mathematical Studies* (Vol. 2, pp. 1–46). Budapest, Hungary: Janos Bolyai Mathematical Society.

Lux, T. (2009). Rational forecasts or social opinion dynamics? Identification of interaction effects in a business climate survey. *Journal of Economic Behavior and Organization, 72*, 638–655.

Merton, R. K. (1995). The Thomas theorem and the Matthew effect. *Social Forces, 74*(2), 379–422.

Meyer, C. D. (2000). *Matrix analysis and applied linear algebra*, SIAM, Society for Industrial and Applied Mathematics.

Meyn, S. P., & Tweedie, R. L. (2005). *Markov chains and stochastic stability*. Berlin; Heidelberg: Springer-Verlag.

Moussaïd, M., Kämmer, J. E., Analytis, P., & Neth, P. P. H. (2013). Social influence and the collective dynamics of opinion formation. *PLoS ONE, 8*(11), e78433.

Muoneke, N. (1987). On the stochastic powers of nonnegative reducible matrices. *Linear Algebra and Its Applications, 90*, 57–63.

Ozaki, R., & Sevastyanova, K. (2011). Going hybrid: An analysis of consumer purchase motivations. *Energy Policy, 39*(5), 2217–2227.

Raafat, R. M., Chater, N., & Frith, C. (2009). Herding in humans. *Trends in Cognitive Sciences, 13*, 420–428.

Rasouli, S., & Timmermans, H. (2016). Influence of social networks on latent choices of electric cars: A mixed logit specification using experimental design data. *Networks and Spatial Economics, 16*(1), 99–130.

Rogers, B., & Murillo, D. (2009). Control of opinions in an ideologically homogeneous population. In H. Liu, J. Salerno, & M. J. Young (Eds.), *Social computing and behavioral modeling* (pp. 1–9). Boston: Springer.

Rogers, E. M. (2003). *Diffusion of innovations* (5th ed.). New York, NY: The Free Press.

Sharmeen, F., Arentze, T. A., & Timmermans, H. J. P. (2010, July). Modelling the dynamics between social networks and activity-travel behavior: Literature review and research agenda. *Proceedings of the 12th World Conference on Transportation Research*, Lisbon, July 11–15.

Sharmeen, F., Arentze, T. A., & Timmermans, H. J. P. (2014a). An analysis of the dynamics of activity and travel needs in response to social network evolution and life-cycle dynamics: A structural equation model. *Transportation Research A, 59*(1), 159–171.

Sharmeen, F., Arentze, T. A., & Timmermans, H. J. P. (2014b). Dynamics of face to face social interaction frequency: Role of accessibility, urbanization, changes in geographical distance and path dependence. *Journal of Transport Geography, 34*(1), 211–220.

Simon, H. (1971). Designing organizations for an information-rich world. In M. Greenberger (Ed.), *Computers, communications and the public interest* (pp. 37–52). Baltimore, MD: Johns Hopkins Press.

Surowiecki, J. (2005). *The wisdom of crowds*. New York, NY: First Anchor Books Editions.

Timmermans, H., Khademi, E., Parvaneh, Z., Psarra, I., Rasouli, S., Sharmenn, F., & Yang, D. (2014). Dynamics in activity travel behavior: Framework and selected empirical evidence. *Asian Transport Studies, 3*(1), 1–24.

Wolf, I., Schröder, T., Neumann, J., & De Haan, G. (2015). Changing minds about electric cars: An empirically grounded agent–based modeling approach. *Technological Forecasting and Social Change, 94*, 269–285.

Wu, F., & Huberman, B. A. (2005). *Social structure and opinion formation* (pp. 1–24). Stanford, CA: Stanford University.

Yaniv, I. (2004). Receiving other people's advice: Influence and benefit. *Organizational Behavior and Human Decision Processes*, 93, 1–13.

Yildiz, E., Acemoglu, D., Ozdaglar, A., Saberi, A., & Scaglione, A. (2011, January). Discrete opinion dynamics with stubborn agents. Submitted to Operations Research manuscript OPRE-2011–01–026.

1   Appendix

In DeGroot (1974), a team of individuals, or a committee or a group of interacting agents, is considered among whom some process of opinion formation takes place. The group may be a small one, for example a group of experts asked by the UN to merge their different assessments on the magnitude of the world population in 2050 into one single judgement. The group might reach agreement on the value of an unknown parameter θ. The unknown parameter θ can be the magnitude of the world population in 2050, the value of an asset, the price of a house, the number of users of some device and so on.

    The problem is to find a consensus, or a shared opinion, within a set of n agents who revise their opinions by averaging (or "pooling") other agent's opinions. In particular, for each pair of agents (i,j), agent i assigns a weight a_{ij} (with $0 \le a_{ij} \le 1$) to the opinion of agent j on the basis of the importance he assigns to the opinion of agent j. If the ith agent feels that the jth agent is a leading expert with regard to predicting the value of the parameter θ or if he thinks that the jth agent has had access to a large amount of information about the value of θ, then agent i will choose a large value for a_{ij}. Alternatively, if agent i feel to have access to much of the information she needs or trusts on herself more than on others, she may wish to assign a large weight a_{ij} to her own opinion and small total weight to the opinion of others (a *self-centred* or *stubborn* or *forceful* agent). The mutual attention of all agents is fully described by the attention matrix $\mathbf{A}$, whose ith row contains the weights of ith agent. Since it is assumed that 100% of the attention of agent i is divided among others and own subjective opinion, the sum of weights for each agent sums up to 1. Thus, the n × n matrix $\mathbf{A}$ is stochastic.

    A vector $\mathbf{F}(k)$ of estimates of parameter θ generates at time k. The ith component of the vector is the opinion about the value of parameter θ of agent i. This is obtained by the previous value $\mathbf{F}(k-1)$ by pooling the opinions of the other individuals within the team by the transition matrix $\mathbf{A}$ through the recurrent condition:

$$F(k) = AF(k-1)$$

Since for $k = 1$, $\mathbf{F}(1) = \mathbf{A}\,\mathbf{F}(0)$; for $k = 2$, $\mathbf{F}(2) = \mathbf{A}\,\mathbf{F}(1) = \mathbf{A}^2\,\mathbf{F}(0)$), it turns out that for $k \ge 1$,

$$F(k) = A^k F(0)$$

An implicit assumption is that if agent i is informed of the distributions of each of the other members of the group, she might wish to revise her subjective distribution to accommodate this information. Given the initial profile $F(0)$, the behaviour of the system is given by the properties of A^k or, in other words, how the attention of the agents to the others are shaped. The simplest case occurs when A has constant entries, that is attention toward herself and towards other agents is constant and does not vary with time. It is assumed that the members of the group continue making revisions indefinitely or until no further revision is necessary to change any member's subjective distribution. In mathematical terms, consensus is reached if and only if there is a distribution F^*, such that

$$\lim_{k \to \infty} F(k) = F^*$$

Stability is reached when, for a given small $\varepsilon > 0$, the difference in norm between two subsequent distributions, that is between $F(k + 1)$ and $F(k)$ is less than ε. Consensus is not the only possible configuration; polarization or fragmentation may also occur. According to the topology of the network, it may even happen that the system bounces back and forth to two or more different configurations.

4 Social media and travel behaviour

Emmanouil Chaniotakis, Loukas Dimitriou and Constantinos Antoniou

Introduction

Social media (SM) has emerged as a prominent trend in social communication. Online platforms, such as Facebook and Twitter, conquer the Internet with millions of visitors and users per day. For the past few years the rankings of most visited web pages (Alexa.com) illustrate the steady positioning of SM and its overall use. Specifically, in February 2018, Facebook was ranked as the 3rd, Twitter as the 13th and Instagram as the 15th most visited websites. These ranks are widely indicative of the overall trend, especially when we consider that most of the remaining top 15 websites are search engines.

Although the history of SM is short, with their wide use to emerge in the 2000s, the idea of opinion exchange and in general communication platforms is not at all new. The first attempt was recorded in 1979, with Tom Truscott and Jim Ellis creating Usenet, a virtual space for users to post public messages (Kaplan & Haenlein, 2010). Following up on that, Internet Relay Chat (IRC) debuted in 1988 as a way for friends and strangers to communicate with major success and paved the road for modern forum platforms, such as bulletin board systems. Towards this direction, SM started as a space for communication, user profiles and updates with SixDegrees.com, often considered as the first SM platform, launched in 1997 (Boyd & Ellison, 2007).

These initiatives, and developed platforms, were providing the means for sharing (or exchanging) personal information about a variety of topics and, in turn, facilitating the collection of the participants' behaviour (in terms of posts, responses, or just reading) in a very organized manner. Focusing on topics of "everyday life", travel behaviour stands for a significant activity consuming a valuable share of peoples' daily time budget, where decisions are made based on the available information that travellers have, and thus users', to have a strong incentive to search for updated and reliable information. Moreover, based on the fact that SM provides a two-way location- and person-specific communication structure (users are contributing information to the "networked community"), it gradually becomes a rich "bank" of disaggregate behavioural information about users' characteristics, information collection activities, preferences, reactions, choices and other elements that ultimately constitute behaviour.

In particular, these days, SM usage generates an astonishing amount of information (and, consequently, data). Interestingly, most users use their smartphones for accessing SM applications – 88% of Facebook and 83% of Twitter users on mobile (www.pewinternet.org/fact-sheet/social-media). This vast "information bank", contains records about travel activities, which extends typical information that users have access to (network/route conditions, traffic information, etc.), toward a user-centric, both fetch- and push-type, updated information that affects travellers' decisions.

On the other hand, the collection of transportation-related data conventionally relies on either travel surveys or traffic counts. However, both methods have limitations, and the validity of the data collected is frequently questioned. Transportation surveys' execution is costly, and there is a long-standing debate on their ability to capture human behaviour. On the other hand, traffic counts do not offer the data proximity, as well as the spatial coverage, required, especially in urban settings. These limitations have imposed the development of voluminous methods and techniques aiming to accurately represent the transportation system and shape transport policies, using the available means. Some of the limitation of the conventional types of data and data collection methods can be overcome with the evolution of pervasive systems, such as Global Positioning System (GPS) handsets and cellular networks, and particularly of SM, such as Facebook, Twitter, Flickr and Google+. Data originating both directly from pervasive systems and from SM have shifted the interest of the scientific community from the focus on overcoming limitations towards utilizing the use of the increasingly available data, namely Big Data. Specifically, the utilization of user-generated content (UGC), originating from SM, has been examined in the literature. However, the validity of the data, as well as the context of usage, has received little attention.

In this chapter we critically review various components of the use of SM in transport. We approach this topic by first providing a short summary of crucial aspects of travel behaviour that are related to SM use. We continue with the review of the use of SM in transport, discussing how the field has emerged and the directions it has taken, as well as some privacy and data-related implications with regards to the use of social media data. Finally, aiming at showcasing its potential, three use cases that illustrate possible uses of SM are summarized.

Travel behaviour

In order to understand the effect that SM has on transportation systems operations (mainly), it is valuable to recall some fundamental elements of the system organization and integration. For the purposes of the current chapter, among various approaches that could be used, the description of transportation systems organization will follow the socio-economic approach, where social systems are described based on demand–supply inter-relations. Focusing on the term *travel demand* (especially regarding personal travel), this corresponds to manifestation of choices made by all (potential and actual) users of the available transportation supply system (infrastructure), or, in other words, the manifestation of people's behaviour that reflects their characteristics, their needs, their opportunities and their available information. The developments in the information, communications and marketing scientific and commercial fields provided (or promoted) additional opportunities for ad hoc connections among people interested in sharing or exchanging information about particular topics, an element of interaction that – as earlier described – corresponds to a particular type of social behaviour. Also, by considering the fundamental axiom that behaviour is "governed" (beyond personal characteristics) by the available information, it is evident that the (almost-sudden) blossom of the additional channel of information that SM offered; inevitably, SM are playing an increasingly vital role in peoples' choices (and thus behaviour). Finally, as much as travel activities are having an important role in everyday life, these are affected by information recorded in SM that is relevant, shared by the connected community and can be reached (and appraised as valuable) by potential users/travellers.

On the other hand, the development of a communication network of interested users provided an additional mean from the side of system operators in order to improve systems' performance, in an intervening scheme. In particular, the fact that information provided may influence the users' choices (and thus travel demand), the typical scheme used in transportation systems of 'guiding' users' choices may be employed using the opportunities of SM. As such, suitably selected/composed "messages" could be broadcasted, such as to encourage (discourage) users of making particular decisions, resulting to a targeted, timely and effective additional mechanism for treating travel demand and optimize system performance. In order to achieve such objectives, a thorough multidisciplinary analysis is needed. Possibly new methodological tools should be employed or developed and additional scientific disciplines should be engaged in order first understand the mechanisms of this new type of communication channel (and data sets available), and then optimal intervention strategies can be formed. In the following paragraphs, a brief description of the use of SM data in transportation is provided.

SM in transport

SM characteristics have been the subject of multiple studies. Their particular focus in terms of functionalities that each SM covers can be organized in seven blocks (Kietzmann, Hermkens, McCarthy, & Silvestre, 2011).

SM functionalities

Chaniotakis, Antoniou and Pereira (2016) have adjusted these seven blocks to reflect the impact that SM has on transportation. In particular, (a) **Presence** describes when someone is accessible and/or located. This functionality, and especially its reflection on the real world, is interesting in transportation research, because it lets researchers analyse traces from individuals to study mobility and activity patterns. (b) **Sharing** refers to content that can be shared. It is interesting to mention that some platforms are built around a particular type of content, whereas others embed content within other functionalities. SM content-sharing exploration can provide valuable information on the individual identity and on some of the personality characteristics. However, such an endeavour would require the use of advanced image- and video-processing algorithms and would raise privacy issues. (c) **Relationships** refers to the inclusion of social networks (defined as a network of social interactions and personal relationships). The exploration of online social network structures and interactions can contribute to the exploration of the impact that social networks have to mobility patterns (Axhausen, 2008). (d) **Identity** describes the extent to which SM allows and requires users to reveal their true identity by including information and can be perceived as the act of disclosing thoughts and feelings that describe an individual's preferences (Kaplan & Haenlein, 2010). Identity characteristics are particularly interesting for transportation research on the identification of samples for modelling purposes. (e) **Interactions** describes the communication (via messages, pokes, and posts) with people in their SM network and with strangers (in some cases, from groups defined within each SM platform). SM conversations can be performed in public or with private messages. The interactions that take place among individuals in SM platforms and their tone, form and content can reveal each individual's social network and, consequently, can allow transportation researchers to get insights in the social network and mobility relation. They also allow for the identification of identity and personality characteristics that have been found to be factors affecting mobility. (f) **Groups** refers to the formation of communities and groups. (g) **Reputation** refers to the individual and the content posted. Reputation is particularly important in transport as it describes the scheme of internet influencers, people who are being followed in order to provide advice for, in many cases, transport-related activities.

SM use in transport

Data from SM can be categorized as a passively collecting data source, forming a particularly useful user-generated data source. Efforts in working with this yet growing amount of data have been directed towards all aspects of the Big Data life cycle (data acquisition, information extraction and cleaning, data integration, aggregation and representation, modelling analysis and interpretation; see Jagadish et al., 2014), constituting a rather multidisciplinary research topic. In transportation, these efforts have been mainly focusing on the aspects of data acquisition – mostly in terms of data collection – information extraction and cleaning and modelling analysis.

The literature on SM and travel behaviour is vast. Rashidi, Abbasi, Maghrebi, Hasan and Waller (2017) presented a discussion on the evolution of the literature related to SM and transport. They performed a Scopus search only for title and abstract, which resulted in 935 papers until late 2015 performing the following query:

> ("Social media" OR Twitter OR foursquare OR facebook OR yelp OR instagram) AND ("travel" OR "transport" OR "mobility" OR "geo")

Performing the same search in December 2018 (undertaken by the authors of this chapter), yields 2,442 documents, with approximately 400 papers being published every year. To get a better idea of the various focus areas of social media use in transport, we have performed a number of different queries that illustrate the various research trends that emerge (Table 4.1). For comparison purposes, we have used the same query structure for SM (("Social media" OR Twitter OR foursquare OR facebook OR yelp OR instagram)) and we have been changing the transport-related part.

The commonly exploited SM-originated information is based on the use of the spatial information accompanying posts (geotag) and, in some cases, the language processing of posted content. This is primarily performed in terms of historical data analysis and corresponds to the investigation of research areas, such as travel demand (Lee, Gao, & Goulias, 2016; Maheswaran, Tang, & Ghunaim, 2007), the exploration of activity modelling (e.g. Chen, Mahmassani, & Frei, 2017; Lee, Davis, & Goulias, 2016), the identification of urban settings (e.g. Jiang & Miao, 2015), mobility patterns (e.g. Hawelka et al., 2013) and social networks (e.g. Cho, Myers, & Leskovec, 2011). Also interesting is the use of SM for continuous streaming of information for identifying disruptions or special events and for forecasting (e.g. Gu, Qian, & Chen, 2016; Kumar, Jiang, & Fang, 2014; Marcus et al., 2011). Not limited to the use of data, SM exploitation for transportation purposes is reflected in the direct communication of transport-service providers, with users and information sharing by SM platforms (Gal-Tzur, Grant-Muller, Minkov, & Nocera, 2014). For more details on the different areas of using SM for transportation studies, the reader can refer to Chaniotakis et al. (2016). In the next few paragraphs, we briefly explore the use of social media for the categories identified as most pertinent to travel behaviour.

Interestingly, different trends emerge with the establishment of SM research in transport. As presented in Figure 4.1, although in the first few years, the keyword "traffic" was dominant, there has been a gradual but steady swift towards "tourism" and "travel".

Table 4.1 Number of papers resulting from Scopus search based on keywords used

Keyword[*]	*Total Number of Papers*
Traffic	1096
Travel	838
Tourism	791
Mobility	685
Transportation	403
Trip	227
Travel behaviour	44
Travel demand	21

[*] *Query performed as a combination of (("Social media" OR Twitter OR foursquare OR facebook OR yelp OR instagram)) and the keyword mentioned, only for title and abstract*

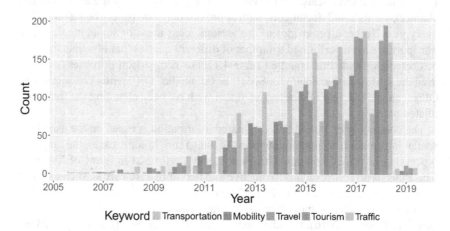

Figure 4.1 Evolution of literature for different keywords explored, based on Scopus search

To summarize, the main interest – especially in the first few years of the exploration of SM use – in transport was around the data properties itself and the resulting travel characteristics. This was explored on the basis of what mobility patterns can be extracted from SM, both on a local (Cheng, Caverlee, Lee, & Sui, 2011) and a global level (Hawelka et al., 2013), focusing, in many cases, to model locations visited (Cho, Ver Steeg, & Galstyan, 2014; Yuan, Cong, Ma, Sun, & Thalmann, 2013). Accordingly, the investigation of the potential to substitute or complement existing data collection methods was explored. While initial efforts aimed to substitute conventional travel survey data, this was quickly abandoned as discrepancies were observed, such as a high representation of specific fractions of activities (e.g. leisure) or differences in direct comparison emerged (Lee, Gao, et al., 2016; Rashidi et al., 2017). Lately, efforts on the use of SM are slowly directed towards the combination of different data sources employing data-fusion methods (Akbar et al., 2018).

Privacy and data implications of SM use

Privacy implications of SM have long been examined by different angles. Two main streams can be identified: (a) potential of privacy infringements through social media and (b) changes of privacy perceptions through SM. Both are widely connected to transportation research and the use of social media in transport.

Starting with the latter, it has been evidenced that SM has changed the culture of sharing information, especially for younger generations (Madden et al., 2013). Users are more likely to reveal their real names, talk about their interests and be open in revealing their thoughts and experiences. The reason why this is important for transport is that it relates to the difficulties associated with data collection.

Users revealing true information could lower the value of privacy (Antoniou & Polydoropoulou, 2014), increasing the response rate on transportation-related surveys and allowing for the combination of SM (which provides an inexpensive stream of information) with other sources of data.

On the potential of privacy infringements through SM, the case of Cambridge Analytica (Greenfield, 2018) was one of the cases that changed the landscape on SM privacy awareness and the need of safeguarding user privacy. One could argue that this was only the tip of the iceberg. Users providing geolocated information or information concerning activities could have an adverse impact on their safety and their ability to safeguard their interests, for example insurance premiums and employment (Sánchez Abril, Levin, & Del Riego, 2012). Although numerous measures have been taken from SM providers to safeguard user privacy, the possibility of privacy infringement is more relevant than ever, as new methods of user identification collection of user data emerge (Smith, Szongott, Henne, & Von Voigt, 2012; Zheleva & Getoor, 2009). Users of SM, in many cases, stand helpless against the identification of possible threats and issues, reducing trust and, overall, creating problems, even to a healthy use of SM.

With regards to the use of data from SM, there is a long-lasting discussion on the premises of data provision and inherent biases. Data provision is tightly connected to issues of privacy and the code of contact of using sensitive data. As extensively discussed in Chaniotakis et al. (2016), data availability relies on the SM platform providers' policies, and one of the main disadvantages of its use is the uncertainty with regards to the continuation of its provision. Already in the last few years, and due to privacy concerns, many providers changed the policies concerning the use of Application Programming Interfaces and limited the data provided (e.g. provision of the number of visits from Foursquare, events-visited data from Facebook) in ways that could restrict its use in transport. With regards to biases, the main point raised with the sampling bias that related to the socio-demographics of SM users. With the establishment of SM platforms, this seems to be alleviated (Rashidi et al., 2017); however, there is still the issue of why and when people use SM, which poses the question of the capabilities that SM data have to increase the observability of the transportation system.

Case studies of SM use in transport

Aiming at illustrating the way that SM interacts with travel behaviour, in the next few sections, the exploration of three related case studies takes place. The first case study aims at illustrating how SM platforms can influence travel behaviour by the exploration of the activities shared within them; the second case study aims at illustrating the focus of SM usage in a spatio-temporal pattern, while the third explores the differences of social media use in different areas around the world. Naturally, there are many other applications, focusing on various modes of transport. Arguably, of particular interest are applications related to public transport (e.g. Steiger, Ellersiek, & Zipf, 2014; Cottrill et al., 2017).

Extracting activities from social media

In this case study the implementation of the framework derived by Chaniotakis, Antoniou, Aifadopoulou, and Dimitriou (2017) has been used to combine data from Foursquare and Twitter in order to perform text modelling for the identification of users' activities (Figure 4.2). An efficient method is implemented that considers the locations that a user visits multiple times in order to extract texts that refer to the same activity. User data is classified using a density-based spatial classification (DBSCAN). This allows the distinction of locations visited frequently and enriches the text available for modelling using Foursquare posts, in a robust and fast way. In terms of data collection, Twitter data was collected for a period of one year for the greater London area, resulting in 482,883 unique users and approx. 4.5 million tweets. 90,000 users were selected as a sub-sample to collect 200 tweets (from their timeline). In total, the database included 11,060,814 tweets. From those, 8,141,996 were tweeted in the greater London area. From the geotagged tweets in the greater London area, 3,764,230 (46.2%) included a link that could be parsed, 220,118 of which originated from Foursquare (2.7%).

The Foursquare links were used to extract information concerning the activities that relate to the tweets. As presented in Figure 4.3, a tendency was observed towards leisure activities for all days of a week. Activities such as education and work were clearly represented more highly during weekdays and less represented during weekend days. The activities with the highest representation were found to be the "Bar – Pub" and "Restaurant" activities. Classification methods were applied on the resulting dataset, after removing stop words and punctuation, with the highest accuracy to be observed with the use of Generalized Linear Model via Penalized Maximum Likelihood (GLM) which had an overall classification accuracy of 83%.

The particular case study provided an understanding of the direct merits in decoding social media data and particularly the exploration of users' activities commonly under-represented in conventional travel survey. User activity patterns can also be derived, and the modelling of long-term activity chains can be realized, as SM users share information in a much larger span of time in comparison to conventional surveys.

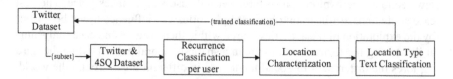

Figure 4.2 User-centric activity enrichment high-level methodology for extraction of activities

Source: Adapted from Chaniotakis et al. (2017).

Spatio-temporal patterns of SM usage

The second case study presents the analysis performed in order to compare the characteristics of SM use in comparison to conventional measurements (Chaniotakis, Antoniou, Grau, & Dimitriou, 2016) on a temporal, spatial and activity level. The main driver of this study is the exploration of the possibilities of using social media in transportation modelling with regards to travel behaviour.

The study was based on the collection of publicly available data from three SM platforms (Facebook, Foursquare and Twitter), which was later compared to a conventionally collected travel survey performed in the same period for the city of Thessaloniki, Greece. Within this case study, the activity characteristics are explored, aiming at identifying which activities are commonly represented by each data source. Then, the temporal characteristics are investigated, including the derivation of correlations among the various data sources. Finally, the spatial distribution of the data is examined in the Thessaloniki city centre using spatial heat maps that extract the density of locations visited from each data source.

On a spatial level, it was clearly evidenced that the distribution of the locations from SM was found to be distributed in areas with a high concentration of recreational land uses which was not observable for the case of the conventional travel surveys, data of which illustrates a more evenly distributed concentration of attractions with little fluctuations in the city centre. This was observable also from the correlations of the observed temporal distributions, where the check-ins for Facebook and Twitter with the travel survey illustrated very low, and in many cases negative, correlations.

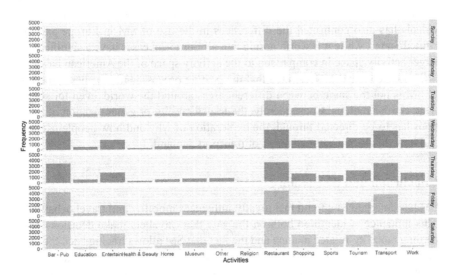

Figure 4.3 Temporal distribution of Foursquare-labelled activities

Source: Adapted from Chaniotakis et al. (2017).

SM use – comparability and transferability

In the third case study, data collected from a selection of 10 cities around Europe and the United States are analysed for the extraction of information to compare different SM use around the world and transferability (Chaniotakis et al., 2017). The different descriptive statistics that can be identified from the REST API data collection of Twitter are first explored, and the timeline (Twitter history) of a sample of users per city is collected. The analysis performed includes the classification of users' posts and the use of distributions to compare different cities' results, following a user-centric analysis of the posting activity and the connection with other Social Media Platforms. Additionally, aspects of activity space have been analysed and compared.

The data used was collected for four cities in the United States (Los Angeles, New York, Orlando, Seattle) and six cities in Europe (Amsterdam, Athens, Copenhagen, London, Munich, Paris). Based on the collected data set, a random sample of at least 1,000 users were selected for each city in order to collect their Twitter timeline. Based on a spatial, temporal and activity space analysis, we can conclude that there are clear differences in the use of social media for different areas around the world. Starting from the posting characteristics, the percentages of the number of geotagged tweets posted in each case clearly differ, with cities in the United States having a range of geotagged tweets that is higher than that observed in European cities (32.9% to 48.4% in the USA vs 11.7% to 29.2% in Europe). Specifically, the highest percentage of geotagged tweets (to the total number of tweets) is found in New York (73.9%), while the lowest is found in Copenhagen (24.3%). Differences have also been observed on the spatial distribution of tweets, corresponding to the different urban structures observed between the two continents. Finally, the exploration of the activity space of different individuals has also confirmed the differences in the use of SM in Europe and the United States, while it is worth noting that all European cases have a significantly larger activity space in comparison to the activity space of the American cases.

This case study has illustrated that the posting process, and in general, the use of SM is not the same between different areas around the world, even for somewhat similar regions. Consequently, the transferability or generalization of solutions has to be directed through the exploration of why and how people use SM and how this is translated in terms of transport demand.

Conclusion and discussion

SM has emerged as a trend that greatly influences mobility and travel behaviour. This influence is identified both on the way that travellers make decisions concerning mobility-related matters and the fact that SM allows tracing back the way that these decisions were made and allow for the collection of data that related to understanding these decisions. This chapter offers a comprehensive "guided tour" on the added value that SM is expected to offer in the organization transportation systems, starting from pointing out the specific elements that this new communication paradigm extends the information that traditionally was used in analysing the phenomenon of travel.

The literature review presented highlights advances and methods that have guided the use of SM in transportation research. The most significant outcome is the shift from the exploration of the data properties towards the investigation of its potential to substitute or complement conventional transport data. In this regard, particular interest is paid to identifying interdependencies between SM and travel behaviour by presenting a collection of different use cases and methods. The indicative research outcomes presented were selected mainly in order highlight the degree of detail that SM data sets offer, and that never before were available, for the understanding the transportation system operations, especially that of travel of people. The differences in social media use in the United States and Europe is first discussed, aiming at a better understanding of aspects of SM use and transferability of solutions. Then, the potential of using the posted text is investigated for the exploration of activities; this is believed to be one of the most valuable information that SM data can provide in this context. Finally, the direct comparison between different sources of data highlights the need for using heterogeneous data in transport-related applications.

The collection of SM data is a process that is amenable to a great degree of automation. Kuflik et al. (2017) describe the potential and challenges of automating this process. It should be noted, however, that besides the merits of using SM in transportation studies, the highlighted issues concerning privacy and data availability can force the use of poorer – in terms of information – data sets that may rely on extremely expensive data collection processes of small sample sizes. The same is evidenced with other (big) data owners that – although collecting data of high value for transportation – refuse to open it for research.

References

Akbar, A., Kousiouris, G., Pervaiz, H., Sancho, J., Ta-Shma, P., Carrez, F., & Moessner, K. (2018). Real-time probabilistic data fusion for large-scale IoT applications. *IEEE Access, 6*, 10015–10027. https://doi.org/10.1109/ACCESS.2018.2804623

Antoniou, C., & Polydoropoulou, A. (2014). The value of privacy: Evidence from the use of mobile devices for traveler information systems. *Journal of Intelligent Transportation Systems* https://doi.org/10.1080/15472450.2014.936284

Axhausen, K. W. (2008). Social networks, mobility biographies, and travel: Survey challenges. *Environment and Planning B: Urban Analytics and City Science, 35*, 981.

Boyd, D. M., & Ellison, N. B. (2007). Social network sites: Definition, history, and scholarship. *Journal of Computer-Mediated Communication, 13*, 210–230. https://doi.org/10.1111/j.1083-6101.2007.00393.x

Chaniotakis, E., Antoniou, C., Aifadopoulou, G., & Dimitriou, L. (2017). Inferring activities from social media data. *Transportation Research Records Journal of Transportation Research Board, 2666*, 29–37. https://doi.org/10.3141/2666-04

Chaniotakis, E., Antoniou, C., & Goulias, K. (2017). Transferability and sample specification for social media data: A comparative analysis. In International Conference on Intelligent Transport Systems in Theory and Practice, Mobil.TUM., Munich, Germany.

Chaniotakis, E., Antoniou, C., Grau, J. M. S., & Dimitriou, L. (2016). Can social media data augment travel demand survey data? In IEEE Conference on Intelligent Transportation Systems, Proceedings, ITSC. https://doi.org/10.1109/ITSC.2016.7795778

Chaniotakis, E., Antoniou, C., & Pereira, F. C. (2016). Mapping social media for transportation studies. *IEEE Intelligent Systems*. https://doi.org/10.1109/MIS.2016.98

Chen, Y., Mahmassani, H. S., & Frei, A. (2017). Incorporating social media in travel and activity choice models: Conceptual framework and exploratory analysis. *International Journal of Urban Sciences*, 1–21. https://doi.org/10.1080/12265934.2017.1331749

Cheng, Z., Caverlee, J., Lee, K., & Sui, D. Z. (2011, July). Exploring millions of footprints in location sharing services. In *Fifth International AAAI Conference on Weblogs and Social Media*. Barcelona, Spain.

Cho, E., Myers, S. A., & Leskovec, J. (2011). Friendship and mobility: User movement in location-based social networks. In *Proceedings of the 17th ACM SIGKDD International Conference on Knowledge Discovery and Data Mining*, KDD '11 (pp. 1082–1090). New York, NY: ACM. https://doi.org/10.1145/2020408.2020579

Cho, Y.-S., Ver Steeg, G., & Galstyan, A. (2014). Where and why users 'Check in'. Proceedings of the Twenty-Eighth AAAI Conference on Artificial Intelligence July 27–31, 2014, Québec City, Québec, Canada (pp. 269–275).

Cottrill, C., Gault, P., Yeboah, G., Nelson, J. D., Anable, J., & Budd, T. (2017, April). Tweeting transit: An examination of social media strategies for transport information management during a large event. *Transportation Research Part C: Emerging Technologies*, *77*, 421–432. https://doi.org/10.1016/j.trc.2017.02.008

Gal-Tzur, A., Grant-Muller, S. M., Minkov, E., & Nocera, S. (2014). The impact of social media usage on transport policy: Issues, challenges and recommendations. *Procedia – Social and Behavioral Sciences*, *111*, 937–946. https://doi.org/10.1016/j.sbspro.2014.01.128

Greenfield, P. (2018). The Cambridge Analytica files: The story so far [WWW Document].

Gu, Y., Qian, Z. (Sean), & Chen, F. (2016). From Twitter to detector: Real-time traffic incident detection using social media data. *Transportation Research Part C Emerging Technologies*, *67*, 321–342. https://doi.org/http://dx.doi.org/10.1016/j.trc.2016.02.011

Hawelka, B., Sitko, I., Beinat, E., Sobolevsky, S., Kazakopoulos, P., & Ratti, C. (2013). Geo-located Twitter as the proxy for global mobility patterns. *Cartography and Geographic Information Science*, *41*(3), 260–271. doi: 10.1080/15230406.2014.890072.

Jagadish, H. V., Gehrke, J., Labrinidis, A., Papakonstantinou, Y., Patel, J. M., Ramakrishnan, R., & Shahabi, C. (2014). Big data and its technical challenges. *Communications of the ACM*, *57*, 86–94.

Jiang, B., & Miao, Y. (2015). The evolution of natural cities from the perspective of location-based social media. *The Professional Geographer*, *67*, 295–306.

Kaplan, A. M., & Haenlein, M. (2010). Users of the world, unite! The challenges and opportunities of social media. *Business Horizons*, *53*, 59–68. https://doi.org/http://dx.doi.org/10.1016/j.bushor.2009.09.003

Kietzmann, J. H., Hermkens, K., McCarthy, I. P., & Silvestre, B. S. (2011). Social media? Get serious! Understanding the functional building blocks of social media. *Business Horizons*, *54*, 241–251. https://doi.org/10.1016/j.bushor.2011.01.005

Kuflik, T., Minkov, E., Nocera, S., Grant-Muller, S., Gal-Tzur, A., & Shoor, I. (2017). Automating a framework to extract and analyse transport related social media content: The potential and the challenges. *Transportation Research Part C: Emerging Technologies*, *77*, 275–291. ISSN 0968–090X, https://doi.org/10.1016/j.trc.2017.02.003

Kumar, A., Jiang, M., & Fang, Y. (2014). Where not to go? Detecting road hazards using twitter. *Proceedings of the 37th international ACM SIGIR conference on Research & development in information retrieval.* ACM, 2014 (pp. 1223–1226). https://doi.org/10.1145/2600428.2609550

Lee, J. H., Davis, A. W., & Goulias, K. G. (2016). Activity space estimation with longitudinal observations of social media data. In Transportation Research Board 95th Annual Meeting. Washington, DC.

Lee, J. H., Gao, S., & Goulias, K. G. (2016). Comparing the origin-destination matrices from travel demand model and social media data. In Transportation Research Board 95th Annual Meeting.

Madden, M., Lenhart, A., Cortesi, S., Gasser, U., Duggan, M., Smith, A., & Beaton, M. (2013). Teens, social media, and privacy. Pew Research Center, 21 (pp. 2–86).

Maheswaran, M., Tang, H. C., & Ghunaim, A. (2007). Towards a gravity-based trust model for social networking systems. In *27th International Conference on Distributed Computing Systems Workshops (ICDCSW'07)*, Toronto, Ontario (pp. 24–24). https://doi.org/10.1109/ICDCSW.2007.82

Marcus, A., Bernstein, M. S., Badar, O., Karger, D. R., Madden, S., & Miller, R. C. (2011). TwitInfo: Aggregating and visualizing microblogs for event exploration. *Proceedings of the SIGCHI Conference on Human Factors in Computing Systems*, (pp. 227–236). ACM. https://doi.org/10.1145/1978942.1978975

Rashidi, T. H., Abbasi, A., Maghrebi, M., Hasan, S., & Waller, T. S. (2017). Exploring the capacity of social media data for modelling travel behaviour: Opportunities and challenges. *Transportation Research Part C: Emerging Technologies, 75*, 197–211.

Sánchez Abril, P., Levin, A., & Del Riego, A. (2012). Blurred boundaries: Social media privacy and the twenty-first-century employee. *American Business Law Journal, 49*, 63–124.

Smith, M., Szongott, C., Henne, B., & Von Voigt, G. (2012). Big data privacy issues in public social media. In Digital Ecosystems Technologies (DEST), 2012, 6th IEEE International Conference on IEEE, Campione d'Italia, 2012, (pp. 1–6).

Steiger, E., Ellersiek, T., & Zipf, A. (2014, November). Explorative public transport flow analysis from uncertain social media data. In *Proceedings of the 3rd ACM SIGSPATIAL International Workshop on Crowdsourced and Volunteered Geographic Information*, (pp. 1–6), Campione d'Italia ACM.

Yuan, Q., Cong, G., Ma, Z., Sun, A., & Thalmann, N. M. (2013). Who, where, when and what: Discover spatio-temporal topics for twitter users. In *Proceedings of the 19th ACM SIGKDD International Conference on Knowledge Discovery and Data Mining* (pp. 605–613). Chicago, IL: ACM.

Zheleva, E., & Getoor, L. (2009). To join or not to join: The illusion of privacy in social networks with mixed public and private user profiles. In *Proceedings of the 18th International Conference on World Wide Web*, WWW '09 (pp. 531–540). New York, NY: ACM. https://doi.org/10.1145/1526709.1526781

5 The relationship between social networks and spatial mobility

A mobile phone–based study in Estonia

Anniki Puura, Siiri Silm and Rein Ahas

Introduction

Many researchers have highlighted the importance of social networks based on interpersonal relationships in studies of social phenomena (e.g. Castells, 1996). Social networks influence all spheres of human activity; furthermore, the networking of various societal phenomena can be noted in present-day society (Castells, 2010). This means that unlike earlier socially and spatially limited communities, nowadays many systems function as increasingly more complex networks.

Researchers of transport and travel behaviour are also paying greater attention to the influence of social networks and communication in travel behaviour and the use of transportation (Carrasco & Miller, 2006; van den Berg, Arentze, & Timmermans, 2012). The influences tend to be one-way – the size of the social network and communication activity have a positive effect on mobility; those who communicate frequently and with a large number of people are also physically the most mobile (Carrasco, Hogan, Wellman, & Miller, 2008; Tillema, Dijst & Schwanen, 2010; van den Berg et al., 2012; Yuan, Raubal, & Liu, 2012). However, the relationships between social networks and spatial mobility are becoming more complex due to two factors. First, the increase in spatial mobility is affected by the relaxing of international border controls and the rapid development of transport infrastructure. Second, social networks, mobility and transportation systems are influenced by developments in information and communications technology (ICT). Many studies (Mascheroni, 2007; Tillema et al., 2010) indicate that communication activity and social networks, and as a result spatial mobility, have increased because of ICT. At the same time, the substitution effect also plays its part – ICT is replacing physical mobility in certain types of communication and activities (Mokhtarian, Salomon, & Handy, 2006; Cooke, 2013). How ICT-based communication affects people's spatial mobility, and whether ICT decreases or increases travel demand is a central topic in spatial planning, geography and transport economics (Line, Jain, & Lyons, 2011; Aguiléra, Guillot, & Rallet, 2012).

Despite the importance of the topic, rather few studies have been published on the inter-relationships between social networks and spatial mobility. There are many reasons for this. First, the changes in society and technology have been extremely rapid in recent years, and a research agenda is still under development. Second, this interdisciplinary field has a number of field-specific approaches (transportation studies, sociology, physics, geography, mathematics, logistics, etc.), and a common theoretical and methodological framework has not yet been established. Third, there is insufficient suitable statistical data in national research and statistical databases. Data revealing the relationships between social networks and spatial mobility at the individual level are even more sparse. As such, one strand of contemporary studies has explored ICT-based resources known as "big data." Social media (Heidemann, Klier, & Probst, 2012; Takhteyev, Gruzd, & Wellman, 2012; Kane, Alavi, Labianca, & Borgatti, 2014) and mobile phone–based data (Calabrese, Smoreda, Blondel, & Ratti, 2011; Phithakkitnukoon, Smoreda, & Olivier, 2012; Yuan et al., 2012; Wang, Kang, Liu, & Andris, 2015; Shi, Wu, Chi, & Liu, 2016) have been used most frequently. Both allow digital connections to be made between a person's communication, mobility, and other characteristics.

The aim of this article is to evaluate the inter-relationships between networks of mobile phone–based calling partners (part of social network) and spatial mobility. In this study, data from Estonian mobile-phone call graphs and from call detail records (CDR) have been used to chart these inter-relationships.

The research questions are as follows:

- How are the size of the networks of calling partners and the locations of the places of residence of the calling partners related to an individual's spatial mobility?
- To what extent is the relationship between the network of calling partners and spatial mobility dependent on an individual's characteristics?

Theoretical framework

Social networks and geography

A social network is a structure composed of people and their relationships, and is based on dyadic relationships between two parties. Traditional social networks are composed of people related to each other via their roles (e.g. family, friends, neighbours, colleagues) (Borgatti, Mehra, Brass, & Labianca, 2009). The relationships may be acquired by birth (family), be derived from membership in a group (school, class) or be from geographic proximity (house, village) (Urry, 2003). Personal suitability is also important in creating and maintaining relationships, as is homophily; that is, people with similar characteristics (age, nationality, etc.) tend to relate better (McPherson, Smith-Lovin, & Cook, 2001).

Three important geographical factors can be highlighted in the functioning of social networks. First, geographical proximity is important in creating and maintaining a number of networks via families, neighbours, schoolmates or those engaged in common activities (Larsen, Urry, & Axhausen, 2006; Borgatti et al., 2009, Sharmeen, Arentze, & Timmermans, 2015). Second, an important characteristic of close social networks is face-to-face meetings of network members, which also requires spatial proximity (Borgatti et al., 2009; Yousuf & Backer, 2015). Third, although certain networks (e.g. families and close friends) may be geographically dispersed, their durability is determined by various relationships (ownership, interests, a sense of closeness, etc.) and face-to-face meetings (Viry, 2012; Carrasco & Miller, 2006; Carrasco, Miller, & Wellman, 2008; Larsen et al., 2006).

The distance decay effect tends to apply to social networks – that is the greater the distance, the more energy it requires to create and maintain relationships, such that fewer network members live farther away from it (Kowald et al., 2013; Calabrese et al., 2011). Both physical and electronic communication frequency decreases with increasing distance (Tillema et al., 2010; Mok, Wellman, & Carrasco, 2010), as does the likelihood of choosing electronic communication over face-to-face meetings (van den Berg et al., 2012). Mobile-based network studies have revealed that the probability and the number of phone calls decrease as physical distance increases (Onnela, Arbesman, Gonzalez, Barabasi, & Christakis, 2011; Lambiotte et al., 2008). The effect of distance on interactions is also well represented in previous studies that use community detection methods (Expert, Evans, Blondel, & Lambiotte, 2011; Ratti et al., 2010). Most of the call activities in these studies take place inside the same administrative units. In addition, some administrative units tend to be more connected than others, reflecting functional relationships between regions. A clear effect of language barriers on interactions can also be seen (Expert et al., 2011).

Activity space and spatial mobility

Activity space refers to places visited as well as to travel to and around those places (Golledge & Stimson, 1997; Dijst, 1999). The part of the activity space used every day is referred to as the "daily activity space"; in tourism and in statistical terms, this is also called the "usual environment" (OECD, 2016; Eurostat, 2016a). The activity space consists of regularly, less regularly and randomly visited places. Regularly visited places are referred to as "anchor points"; these are related to an individual's frequent activities and are places where an individual spends a considerable amount of time and are relevant from the perspective of a person's everyday life (Flamm & Kaufmann, 2006). The most important regularly visited places are the home and the workplace, but shops, gyms, homes of friends and so forth are also important (Golledge & Stimson, 1997; Schönfelder & Axhausen, 2003).

An activity space is measured in two ways: first, by using a point-based approach to evaluate the indicators related to the places visited and the degree of mobility between these locations and, second, using spatial statistics and geometric shapes (Schönfelder & Axhausen, 2010). The point-based approach shows the geography and frequency of the visited places. Geometric shapes enable measurement of the size and characteristics of the activity space; most frequently used are confidence ellipses, standard deviational ellipses (Dijst, 1999; Schönfelder & Axhausen, 2010; Järv, Ahas, & Witlox, 2014) and minimum convex polygons (Buliung, Roorda, & Remmel, 2008; Kamruzzaman & Hine, 2012). A large majority of the extent, geometry, and geography of a person's activity space is found to be determined by his or her most important activity locations (e.g. place of residence and workplace) around which daily life is concentrated (Golledge & Stimson, 1997; Dijst, 1999; Schönfelder & Axhausen, 2010). Similar to networks, the "distance decay effect" also affects the use of space, meaning that as the distance from the place of residence increases, the number of places visited decreases (Golledge & Stimson, 1997; Schönfelder & Axhausen, 2010).

Relationships between social networks and activity spaces

Previous studies of the relationships between social networks and activity spaces have mostly been conducted at the individual level, analysing network ties using indicators such as frequency of interactions and shared locations (Calabrese et al., 2011; van den Berg et al., 2012; Shi et al., 2016). Some studies also include aggregated measures of social networks and activity spaces (e.g. Carrasco & Miller, 2006; Tillema et al., 2010; Phithakkitnukoon et al., 2012).

Individuals who are related tend to have more similar and overlapping activity spaces, with more shared locations in both time and space (Calabrese et al., 2011; Wang et al., 2015; Shi et al., 2016). Moreover, at the aggregate level, the majority of activity locations tend to be confined within short distances from the home and work locations of the members of their social network (Phithakkitnukoon et al., 2012).

The frequency with which individuals share the same locations with members of their social networks and have face-to-face meetings with them, however, also depends on the characteristics of the individual's social ties and personal networks (e.g. size and composition). People who are more closely related (e.g. family and close friends) tend to have more social interactions (Tillema et al., 2010; Carrasco & Miller, 2006). Having more interactions is, in turn, also related to having a higher number of shared locations and face-to-face social activities (Calabrese et al., 2012; Shi et al., 2016; van den Berg et al., 2012). At a network level, a greater number of social interactions and activities are conducted by people who have larger social networks and who belong to a number of different groups (Carrasco et al., 2008; Tillema et al., 2010; van den Berg et al., 2012).

From another perspective, the social networks and activity spaces of individuals must also be seen as constantly co-evolving, with personal social networks shaping individuals' activity spaces (Carrasco & Miller, 2006; Tillema et al., 2010) and with any changes in activities and activity locations leading to changes in social relationships and networks (Saramäki et al., 2014; Sharmeen et al., 2015). In this way, social networks can both result in and be the outcome of spatial mobility (Borgatti et al., 2009). Despite the changing nature of social networks, however, close relationships are maintained even when activity locations and space change and individuals are distanced from their previous locations and members of their social networks due to time constraints (Saramäki et al., 2014; Sharmeen et al., 2015). This is also confirmed by results showing that higher residential mobility leads to social networks that are geographically more extensive (Viry, 2012).

Factors influencing social networks and activity spaces

A person's spatio-temporal behaviour is an outcome of the relationship among individual factors, interactions with other individuals and external factors, such as the surrounding environment and social structure (Hägerstrand, 1970; van Acker, Wee, & Witlox, 2010; Kwan & Schwanen, 2016).

The effect of socioeconomic factors (gender, nationality, age, education, income, profession, etc.) on an individual's social networks and spatial behaviour has been widely studied (Palchykov, Kaski, Kertész, Barabasi, & Dunbar, 2012; Wrzus, Hanel, Wagner, & Neyer, 2013; Kamruzzaman & Hine, 2012; Miranda-Moreno, Eluru, Lee-Gosselin, & Kreider, 2012; Silm, Ahas, & Nuga, 2013). For example, studies show differences between men and women regarding the amount of social interaction they have (Carrasco & Miller, 2006; van den Berg et al., 2012; Yuan et al., 2012), and the extent of their activity spaces and spatial behaviours (Silm et al., 2013; Hamilton, 2001; Polk, 1996), with women tending to have lower spatial mobility and being less likely to conduct social activities than men. Ethnic minorities are found to be less mobile and to visit fewer locations than the majority population (Silm & Ahas, 2014).

The size and structure of social networks tend to vary with age, increasing until young adulthood and then starting to decrease (Wrzus et al., 2013). Particular types of social network members (such as co-workers or neighbours) are seen to be important only in specific age ranges, while family networks are found to be stable in size over the course of a lifetime (Wrzus et al., 2013; Sharmeen et al., 2015). Young people are more active in socializing and have higher frequencies of social interactions, but this decreases with age (Carrasco & Miller, 2006; Tillema et al., 2010). In addition to the youngest age cohorts, however, van den Berg et al. (2012) also found that the oldest age groups have higher frequencies of face-to-face contact with members of their social networks. Spatial mobility is also influenced by age and is found to be higher for middle-aged people (Yuan et al., 2012; Masso, Silm, & Ahas, 2017).

More than age, social networks and spatio-temporal behaviour are influenced by a person's life cycle (e.g. marital status, number of children, university study, etc.) (Wrzus et al., 2013; van Acker et al., 2010, Saramäki et al., 2014). People place different expectations on their social relationships and have a varying capacity to contribute to those relationships at different stages of their lives (Wrzus et al., 2013). For instance, single people communicate with their friends more (Tillema et al., 2010), while parents communicate more with members of their networks with similarly aged children (Palchykov et al., 2012). Individuals' life cycles also have a significant influence on their use of space; for example people with partners and children have more social activities near the home with friends and family (Carrasco & Miller, 2006; van den Berg et al., 2013).

Higher levels of education and greater income increase socializing and spatial mobility (Carrasco & Miller, 2006; van den Berg et al., 2012). While, on the other hand, some authors have indicated that those with a higher level of education and people who work more tend to see less of their friends (Tillema et al., 2010). Furthermore, an individual's social network and spatial mobility are also influenced by other factors, such as vehicle ownership (Schönfelder & Axhausen, 2010; Kamruzzaman & Hine, 2012), factors related to place of residence and employment (Dijst, 1999; Kamruzzaman & Hine, 2012) and use of ICT tools, that is mobile phones and the Internet (Miranda-Moreno et al., 2012)

The use of ICT tools and social media has enlarged people's social networks and increased communication within them (Castells, 2010), as well as increasing their geographic reach (Larsen et al., 2006; Castells, 2010). At the same time, low access to ICT and other mobility tools (e.g., different transportation modes) can make it more difficult to keep in contact with members of one's social networks over longer distances and, therefore, leads to more local social relationships (Larsen et al., 2006; Matous, Todo, & Mojo, 2013, Kowald et al., 2013). The longer the time spent living in the same city or neighbourhood, the higher the number of social activities at the local level (Carrasco & Miller, 2006), which implies the development of a more local social network.

Finally, social networks and spatial behaviour are influenced by the characteristics of physical and social space. Different areas provide different opportunities for activities, interactions with other people and movement patterns. For people living in more densely populated areas, their activity locations and the locations of the members of their social networks are more spatially concentrated (Phithakkitnukoon et al., 2012). Attractive and multifunctional residential surroundings encourage people to be active near their homes (Meurs & Haaijer, 2001; Scheiner & Kasper, 2003). On the other hand, a well-developed transport infrastructure enables people to move farther afield and to visit a larger number of places. People who live in sparsely populated areas often face forced mobility because the various activity places available to them are more spread out (Meurs & Haaijer, 2001; Scheiner & Kasper, 2003; van den Berg et al., 2013). Greater distances between home and workplace, for example, also lead to greater spatial mobility (Yuan et al., 2012).

Materials and methods

Data

The present study uses passive mobile-positioning data that are automatically stored in the memory or log files held by mobile phone operators. The data originate from one of the largest mobile-phone operators in Estonia, whose mobile phone network covers 99.9% of the territory of Estonia. Thus, the study area is the whole of Estonia covering a total area of 45,000 km². Positioning accuracy is 100 to 500 m in densely populated areas or in areas with denser networks of roads, and 500 to 5,000 m in more sparsely populated areas (Ahas, Aasa, Roose, Mark, & Silm, 2008). Nearly 94% of the population of Estonia use mobile phones (Eurobarometer, 2013).

Two types of passive mobile positioning data have been used: (a) mobile-call graph data that provide information about networks of calling partners; (b) call detail records (CDR) that enable evaluation of spatial mobility. Mobile-call graph data include identification (ID) of a caller linked to the ID of the calling partner (the person who received the call or message), provided that the receiving phone is within the same mobile phone network. The CDR database contains the locations of call activities (outgoing calls, text messages and multimedia messages). A randomly generated phone-user pseudonymous ID, the time of the call activity (to the second) and the location (network cell), are indicated for each call activity. The user IDs are the same in both databases, and the ID allows these databases to be linked. Pseudonymous IDs guarantee the anonymity of the phone users, meaning that they cannot be connected to a specific person or phone number. Every mobile phone user's gender, year of birth and preferred communication language (Estonian, Russian, English), chosen by the user when signing a contract, were provided to researchers by the mobile phone operator for scientific purposes. CDR data were used to identify the home and work anchor points of each phone user via a special anchor-point model based on the location and timing of call activities (Ahas, Silm, Järv, Saluveer, & Tiru, 2010).

All data were issued by the mobile phone operator solely for scientific use, and ID-based analysis only takes place in a restricted environment. The collection, storage and processing of the data complied with EU directives on handling personal data (European Parliament, 1995) and the protection of privacy in the electronic communications sector (European Parliament, 2002). Separate approval was also obtained from the Estonian Data Protection Inspectorate.

The period of study was February 2013, because the mobile-call graph data were accessible between February 18 and 28, 2013 (11 days), thus dictating February 2013 as the study period for the CDR data as well. The analysis includes 70,536 mobile phone users across Estonia corresponding to the following criteria: (1) aged 19 years and older (people below this age were excluded due to under-representation), (2) user language of Estonian or Russian (those using English were excluded due to under-representation), (3) it was possible to determine home and work anchor points, (4) number of call activities was on average between 2 and 22 per day (users making very few or very many calls were excluded because there were insufficient data to estimate the spatial mobility of people making only a few calls and phones generating numerous calls are often connected to technical systems or service providers), (5) at least 50% of the calls were made to calling partners using the same mobile phone operator (this criterion helps guarantee that the majority of the members of the network of the calling partners were included when calculating network characteristics) and (6) reciprocal call activity could be identified with at least one partner.

The number of people included in the study (70,536) is 6.8% of the population of Estonia aged 19 or over. According to the census of 2011, the distribution of gender and settlement hierarchy of place of residence for those taking part in the study is similar to the overall Estonian population. Of the people included in the study, 52.7% were female while 54.8% of the total Estonian population is female. The distribution of places of residence by settlement hierarchy differs by a maximum of 5.1 percentage points; the difference is greatest in rural areas. Russian speakers (Russians, Ukrainians, Belarusians) and the youngest (19–29 years) and oldest (65+ years) age cohorts are under-represented in the study. Russian speakers compose 30% of the population of Estonia but only 12.4% of those in the study.

Methodology

Social networks were analysed from an egocentric perspective, that is networks originating from a particular individual (Wellman, 1988; Borgatti et al., 2009). A person's network of calling partners is defined in this study as consisting of calling partners with whom one has at least one reciprocal call activity during the 11-day period, that is at least one call activity initiated by each party. The reciprocal call activity requirement helps determine the important partners and reduces the number of more incidental call activities (i.e., call activities related to a service provider). A similar approach – determining networks based on partners who mutually make call activities to each other – has been used before (Onnela et al., 2011; Phithakkitnukoon et al., 2012). Spatial mobility has been analysed using indicators determined according to the locations of the call activities. The characteristics of the networks of calling partners and spatial mobility have been determined in a way that allows comparison between networks and describes both the number and placement of locations. The characteristics used in the analysis are indicated in the following section.

Characteristics of the networks of calling partners

Number of calling partners refers to the number of people an individual has called by mobile phone and who have themselves also made calls to the person of interest. On average, individuals have 6.2 such calling partners.

Number of districts containing a calling partner's home is the number of unique districts that include the place of residence of one or more calling partners of the person of interest. The districts are municipalities and towns within rural areas, together with the urban districts of Tallinn, a total of 238 districts in all. The average area of the districts analysed is 186.3 km², but this varies significantly from a minimum of 1.8 km² to a maximum of 871.6 km². On average, an individual's calling partners live in 3.6 different districts.

Average distance to calling partner's home is the average aerial distance between the home district of the individuals examined and the districts containing a calling partner's homes (in kilometres). On average, this is 26. km.

Characteristics of spatial mobility

Number of visited districts is the number of unique districts from which the person has made at least one call during the study period. Districts are the same as those used to analyse the networks of calling partners. An average of 7.5 different districts are visited each month.

Area of the activity space ellipse is the area of the standard deviational ellipse described by the locations from which each person conducted call activities during the study period. The area represents the smallest possible area in which monthly visited activity locations are found with a probability of 95%. The number of call activities made in each location has also been taken into account as a weighted measure in calculating the ellipse, reflecting the importance of these locations (Dijst, 1999; Schönfelder & Axhausen, 2010; Järv et al., 2014). The average area of the activity space ellipse is 2,079.1 km².

The networks of calling partners and spatial mobility characteristics differ somewhat depending on aspects of an individual's characteristics. The values of all the network characteristics of one's calling partners, as well as the number of visited districts and the area of the activity space ellipse, are larger for men than for women and larger for Estonian speakers than for Russian speakers (the differences are statistically significant, $p < 0.01$; see Table 5.1). The average distance to

Table 5.1 The average values of the characteristics of an individual's calling partners' network and their spatial mobility in relation to their individual characteristics

	No. of calling partners	No. of districts containing a calling partner's home	Average distance to calling partners' home (km)	No. of visited districts	Area of the activity space ellipse (km²)
Total	6.2	3.6	26.2	7.5	2,079
Gender: male	7.5	4.4	27.7	9.2	2,680
female	5.1	3.0	25.0	6.0	1,586
Language: Estonian	6.2	3.7	27.1	7.6	2,165
Russian	4.6	2.4	17.4	5.3	1,095
Age: 19 to 29	4.9	2.9	28.6	7.8	2,196
30 to 39	4.8	2.7	20.5	6.5	1,649
40 to 49	4.8	2.6	21.8	6.3	1,745
50 to 64	4.5	2.7	27.1	6.0	1,757
65+	4.0	2.5	26.9	4.5	1,126
Settlement-system hierarchy level of home district: Capital city	6.2	4.1	23.1	7.8	1,510

	No. of calling partners	No. of districts containing a calling partner's home	Average distance to calling partners' home (km)	No. of visited districts	Area of the activity space ellipse (km²)
Capital city commuting area	6.8	4.4	23.5	9.1	1,663
Regional centre	5.8	2.8	27.3	5.7	2,044
Regional centre commuting area	6.3	3.5	27.1	7.1	2,081
County town	6.1	3.3	28.3	6.8	2,635
County town commuting area	6.2	3.5	28.3	7.3	2,613
Small town	6.0	3.3	25.9	8.0	2,725
Rural areas	6.2	3.5	29.8	8.0	2,803
Distance between home and work (km): 0.0 to 0.9	5.2	3.0	23.6	5.9	1,841
1.0 to 4.9	6.2	3.5	23.7	6.8	1,927
5.0 to 9.9	6.8	4.1	23.5	8.2	1,906
10.0 to 49.9	7.1	4.3	27.6	9.3	2,185
50.0+	6.7	4.3	68.3	13.1	5,343

partners' home is significantly (p < 0.01) less for middle-aged (30–49 years old) people, while spatial mobility characteristics have greater average values for the youngest cohort (19–29 years) and smaller values for the oldest cohort (65+ years; p < 0.01). People living in the capital city or its commuting area and those having workplaces closest to their homes (up to 10 km) have calling partners within a shorter average distance and smaller areas of activity space ellipse compared to those living in lower levels of settlement-system hierarchy and those having workplaces farther away from their homes (p < 0.01).

Spearman's (2-tailed) rank correlation coefficient was used to evaluate the strength of the connections between the network of the calling partners and spatial mobility. The effect of the characteristics of the calling partners' network on spatial mobility was also evaluated using a General Linear Model (GLM):

$$Y_i = \alpha + \sum_{k=1}^{k} \beta_k \chi_{ik} + \epsilon_i,$$

where Y_i is the value of the spatial mobility characteristic (number of visited districts, area of the activity space ellipse) by individuals $I = 1, \ldots, i$; α is a constant; χ_{ik} is the value of the variable for an individual; β_k is the parameter describing the impact of this variable, with k variables; and ϵ is the error term. Six models were constructed altogether looking at different characteristics of the network of the calling partners. Separate models, describing different aspects of the network of the calling partners, were created because the network characteristics could not be included in a single model due to their correlations. The first three estimate the number of visited districts, in terms of (1) the number of calling partners, (2) the number of districts containing a calling partner's home and (3) the average distance to the calling partner's home. Models 4, 5, and 6 estimate the area of the activity space ellipse, specifically looking at different characteristics of the calling partners' network in terms of (4) the number of calling partners, (5) the number of districts containing a calling partner's home and (6) the average distance to the calling partner's home. In addition, individual characteristics (gender, language, age, settlement-system hierarchy level of home district, distance between home and work; see Table 5.1) were also incorporated into each model. The models incorporate data from 39,929 people; the network of calling partners and personal characteristics were known for each individual.

Results

Relationships between the networks of calling partners and spatial mobility

The results indicate that the number of calling partners an individual has and their geographical distribution have a positive correlation with a person's spatial mobility. As the number of calling partners, the number of districts containing a calling partner's home, and the average distance to a calling partner's home increases, so does the individual's mobility. All the characteristics of the networks of calling partners have a positive correlation with the characteristics of spatial mobility; furthermore, these correlations are statistically significant ($p < 0.01$).

The characteristics of the networks of calling partners are fairly strongly correlated with the number of districts visited (see Table 5.2). The number of districts containing a calling partner's home is most strongly correlated with the number of visited districts ($\rho = 0.483$), followed by the number of calling partners ($\rho = 0.420$), and the average distance to the calling partner's home ($\rho = 0.320$). Therefore, having more calling partners living in a greater number of districts that are farther away is related to individuals visiting a greater number of districts, indicating higher mobility.

The area of the activity space ellipse is fairly strongly correlated with the average distance to the calling partner's home ($\rho = 0.366$; see Table 5.2). Weaker correlations are with the number of calling partners ($\rho = 0.278$) and the number of districts where calling partners live ($\rho = 0.285$). In this way, individuals' activity spaces are more likely to cover a more extensive area when their calling partners are geographically more distant, and this is somewhat less related to how many calling partners an individual has and in how many different districts these calling partners live.

The number of calling partners and the number of districts containing a calling partner's home account for most of the variation in the number of visited districts when the network characteristics of the calling partner and the individual are added to the model. The average distance to a calling partner's home, on the contrary, accounts for most of the variation in the area of the activity space ellipse (see Table 5.3). The number of districts containing a calling partner's home describes 13.0% and the number of calling partners 11.0% of the variance in the number of

Table 5.2 Correlations between the characteristics of the calling partners' network and spatial mobility (Spearman's rho two-tailed rank correlation coefficients)

	No. of visited districts	Area of the activity space ellipse (km²)
No. of calling partners	0.420	0.278
No. of districts containing a calling partner's home	0.483	0.285
Average distance to calling partner's home	0.320	0.366

Source: Note: All correlations are statistically significant (p < 0.01).

Table 5.3 Differences in the spatial mobility characteristics by an individual's and calling partners' network characteristics according to GLM

	No. of visited districts			Area of the activity space ellipse		
	No. of calling partners	*No. of districts containing a calling partner's home*	*Average distance to calling partner's home*	*No. of calling partners*	*No. of districts containing a calling partner's home*	*Average distance to calling partner's home*
R²						
Corrected Model	0.247**	0.264**	0.174**	0.084**	0.089**	0.116**
Calling partners' network characteristic (in the table's header)	0.110**	0.130**	0.023**	0.011**	0.016**	0.045**
Gender	0.030**	0.031**	0.038**	0.006**	0.006**	0.009**
Language	0.004**	0.002**	0.006**	0.003**	0.002**	0.002**
Age	0.013**	0.018**	0.020**	0.002**	0.002**	0.004**
Settlement-system hierarchy level of home district	0.039**	0.022**	0.038**	0.014**	0.019**	0.014**
Distance between home and work	0.060**	0.051**	0.033**	0.038**	0.036**	0.016**
B						
Calling partners' network characteristic (in the table's header)	0.513**	0.989**	0.022**	129.023**	273.796**	23.960**

(*Continued*)

Table 5.3 (Continued)

	No. of visited districts			Area of the activity space ellipse		
	No. of calling partners	No. of districts containing a calling partner's home	Average distance to calling partner's home	No. of calling partners	No. of districts containing a calling partner's home	Average distance to calling partner's home
Gender (ref. male)	0	0	0	0	0	0
female	−1.427**	−1.431**	−1.692**	−537.216**	−532.681**	−637.212**
Language (ref. Estonian)	0	0	0	0	0	0
Russian	−0.767**	−0.496**	−0.962**	−543.905**	−459.865**	−446.790**
Age (constant)	−0.039**	−0.046**	−0.052**	−11.373**	−13.080**	−17.574**
Settlement-system hierarchy level of home district (ref. capital city)	0	0	0	0	0	0
Capital city commuting area	0.455**	0.634**	0.613**	−125.656*	−77.579	7.560
Regional centre	−1.890**	−1.178**	−1.890**	536.088**	731.853**	489.631**
Regional centre commuting area	−1.189**	−0.747**	−1.017**	272.766**	391.518**	342.761**
County town	−1.302**	−0.707**	−1.203**	775.487**	937.353**	786.971**
County town commuting area	−1.013**	−0.546**	−0.744**	705.116**	828.131**	783.694**
Small town	−0.188	0.389**	0.038	924.907**	1,079.419**	994.905**
Rural areas	−0.388**	0.095	−0.153*	946.468**	1,073.701**	987.351**
Distance between home and work (km)	0.042**	0.038**	0.034**	27.353**	26.317**	17.795**

Source: Significance level: *$p < 0.05$, **$p < 0.01$

visited districts. The average distance to a calling partner's home describes 4.5% of the variance in the area of the activity space ellipse (see Table 5.3). According to this, the characteristics of a social network remain the most important factors influencing individuals' spatial mobility.

Relationships between the network of calling partners and spatial mobility in relation to individual characteristics

The strength of the relationships between the characteristics of the network of calling partners and spatial mobility are also found to be influenced by individual characteristics. The correlation between the network of calling partners and spatial mobility differs most by gender, where the relationship is stronger for men than for women (see Table 5.4). This means that the network of the calling partners influences spatial mobility for men more than for women.

The difference in the correlation coefficients is greatest with regard to the connection between the number of calling partners and the number of visited districts

Table 5.4 Average correlation coefficients between calling partners' network characteristics and spatial mobility characteristics according to an individual's characteristics

	No. of visited districts			Area of the activity space ellipse		
	No. of calling partners	No. of districts containing a calling partner's home	Average distance to calling partner's home	No. of calling partners	No. of districts containing a calling partner's home	Average distance to calling partner's home
Total	0.420	0.483	0.320	0.278	0.285	0.366
Gender: male	0.462	0.518	0.368	0.282	0.290	0.399
female	0.309	0.389	0.260	0.208	0.221	0.322
Language: Estonian	0.406	0.464	0.296	0.251	0.255	0.344
Russian	0.322	0.439	0.330	0.248	0.255	0.329
Age: 19 to 29	0.310	0.371	0.249	0.180	0.187	0.302
30 to 39	0.292	0.401	0.320	0.205	0.228	0.317
40 to 49	0.301	0.391	0.302	0.241	0.249	0.334
50 to 64	0.264	0.345	0.247	0.216	0.235	0.346
65+	0.236	0.325	0.222	0.183	0.219	0.356
Settlement-system hierarchy level of home district: Capital city	0.425	0.443	0.337	0.280	0.307	0.392
Capital city commuting area	0.476	0.504	0.362	0.260	0.267	0.335
Regional centre	0.427	0.469	0.375	0.344	0.399	0.385
Regional centre commuting area	0.415	0.464	0.355	0.301	0.353	0.351
County town	0.399	0.446	0.305	0.301	0.343	0.307
County town commuting area	0.384	0.417	0.280	0.284	0.319	0.279
Small town	0.389	0.386	0.297	0.252	0.259	0.302

(*Continued*)

Table 5.4 (Continued)

	No. of visited districts			Area of the activity space ellipse		
	No. of calling partners	No. of districts containing a calling partner's home	Average distance to calling partner's home	No. of calling partners	No. of districts containing a calling partner's home	Average distance to calling partner's home
Rural areas	0.385	0.439	0.286	0.246	0.284	0.276
Distance between home and work (km): 0.0 to 0.9	0.345	0.400	0.251	0.253	0.249	0.295
1.0 to 4.9	0.411	0.482	0.311	0.309	0.315	0.349
5.0 to 9.9	0.451	0.493	0.305	0.288	0.273	0.355
10.0 to 49.9	0.436	0.476	0.275	0.245	0.235	0.328
50.0+	0.498	0.488	0.082	0.201	0.210	0.267

Note: All correlations are statistically significant, $p < 0.01$.

(0.153). When it comes to the area of the activity space ellipse and the characteristics of the network of calling partners, the greatest difference between the correlation coefficients is found in the connection between the average distance to calling partner's home and the area of the activity space (0.077).

In the case of language, most of the connections between the characteristics of the network of the calling partners and the characteristics of spatial mobility are stronger for Estonian speakers than they are for Russian speakers (see Table 5.4). An exception is the relationship between the average distance to calling partner's home and the number of visited districts – here, the correlation is stronger for Russian speakers. The numerical difference between the characteristics of the network of calling partners and the characteristics of the spatial mobility of Russian speakers and Estonian speakers is minimal. The greatest difference emerges with regard to the connection between the number of calling partners and the number of visited districts (0.084). According to this, in most cases, the calling partners' network influences the spatial mobility more for Estonian- than for Russian speakers.

No clear trend emerges with regard to the relationships between calling partners' network and spatial mobility characteristics with regard to age. Connections between the characteristics of the network of calling partners and the number of visited districts are stronger among younger people and weaker among older cohorts; however, these differences are quite small (see Table 5.4). No similar trend is found regarding the connection between the characteristics of the network of calling partners and the area of the activity space ellipse.

Turning to the settlement-system hierarchy levels of the individual's home district, the strongest relationships between the network of calling partners and spatial mobility can be noted for people living in higher levels of settlement-system hierarchy (see Table 5.4). This implies that the network of calling partners influences the spatial mobility most for people who live in the capital city commuting area, in the capital city and in regional centres. The correlations between the network of calling partners and spatial mobility are lowest in county town commuting areas, small towns and rural areas.

No clear trend emerges in the correlation between the characteristics of the network of calling partners and the characteristics of spatial mobility as the distance between home and workplace changes (see Table 5.4). The correlations between the characteristics of the network of calling partners and the number of visited districts are somewhat smaller for people who work near home (a distance of less than 1 km) and, with regard to the area of the activity space ellipse, for those working far from home (a distance of greater than 50 km).

Discussion and summary

The positive relationship between social networks and spatial mobility noted in earlier studies (Calabrese et al., 2011; Phithakkitnukoon et al., 2012) was also confirmed by this mobile phone–based study conducted in Estonia. The strongest relationships were found between the number of calling partners and districts containing a calling partner's home and the number of visited districts. The average distance to a calling partner's home is most strongly related to the area of the activity space ellipse. These results indicate that people with larger, and geographically more extensive social networks, are also spatially more mobile.

The previously mentioned results relate to previous knowledge that existing contacts motivate a person to travel in order to maintain his or her important social relationships and to gain social support, well-being and so on (Larsen et al., 2006; Carrasco & Miller, 2006; Tillema et al., 2010; van den Berg et al., 2012). On the other hand, more mobile people have been found to create more social ties on a geographically broader scale resulting from more active daily lifestyles or previous changes of activity locations (e.g. home and work locations; Carrasco et al., 2008; Viry, 2012; Sharmeen et al., 2015). This may lead to "chicken-or-egg" types of questions – does a larger network cause greater spatial mobility, or do more active people acquire a larger network? It is likely that no single answer exists, given that the process of the formation of networks and spatial mobility is highly complex and that the different factors are interconnected.

However, knowing that the process of network transformation is rather long term and often related to different life-cycle events (|Wrzus et al., 2013; Sharmeen et al., 2015) we may assume that data from just 11 days allows us to assess the impact of existing social networks on people's spatial mobility rather than vice versa. To gain knowledge about dynamics in social networks related to spatial mobility, longer time periods should be used, covering different life-cycle events and significant changes in activity spaces (Saramäki et al., 2014; Sharmeen et al., 2015). Therefore, the length of the study period must be considered influential in allowing an exploration of relationships between social networks and spatial mobility through different perspectives.

An important result of our study is that relationships between the networks of calling partners and spatial mobility are influenced by individual characteristics, where the greatest difference is found between men and women, followed by the different levels of hierarchy of settlement systems of home districts. Slight differences are also evident in language (minority vs majority) and age. No clear effect emerges according to the distance between home and workplace.

Mobile phone–based data do not enable the analysis of the reasons behind these differences in greater detail. As for why the connection between characteristics of a network of calling partners and the characteristics of spatial mobility is stronger for men than for women, it might be claimed that men are more spatially mobile (Silm et al., 2013; Hamilton, 2001; Polk, 1996) and that they tend to engage in more social activities than women do (Carrasco & Miller, 2006; van den Berg et al., 2012). Differences between the genders are usually associated with different societal roles, including raising children (Silm et al., 2013; Palchykov et al., 2012). Injustice is another frequently cited cause; for example men and women have different income levels and access to transportation (Cristaldi, 2005; Hamilton, 2001). The difference in income must be highlighted when interpreting the Estonian results – Estonia, after all, has the greatest gender pay gap in the European Union. The average wage differs by 28.3% between men and women in Estonia while the EU average is 16.1% (Eurostat, 2016b). These factors are likely to influence the relationship between social networks and spatial mobility.

Differences in language may refer to segregation in Estonia, where Russian speakers, as a minority, are often segregated, living and functioning in different areas and having more spatially limited activity spaces (Silm & Ahas, 2014). The stronger correlation for younger and middle-aged groups can also be explained by their greater mobility (Yuan et al., 2012) and by the fact that the frequency of social interactions decreases with age (Carrasco & Miller, 2006; Tillema et al., 2010). The overall size of personal social networks also decreases with age, when, for example, contacts related to work may lose relevance (Wrzus et al., 2013).

Location of the home district is found to be important, and the strongest relationships between networks of calling partners and spatial mobility are noted for people living in higher levels of settlement-system hierarchy. In contrast, the weakest relationships are among people living in lower levels (e.g. small towns and rural areas). This may result from higher levels of hierarchy, featuring higher population density and more services and providing more activity locations for social activities (Meurs & Haaijer, 2001; Scheiner and Kasper, 2003; Phithakkitnukoon et al., 2012). At the same time, people living in sparsely populated areas often travel for purposes other than socializing (e.g. services) (Meurs & Haaijer, 2001; Scheiner & Kasper, 2003). The distance between home and workplace may have an effect on the relationships between the characteristics of the network of calling partners and spatial mobility. However, clear trends from this study are not found.

In addition to individual factors, connections between social networks and spatial mobility are also context-related; that is they are influenced by external factors (e.g. the size and geography of the country). For example it is not possible to compare Estonia (with an area of 45,000 km^2) with Canada (9 million km^2) in terms of internal migration. This knowledge is important, especially because the current study is limited to Estonia and does not address social contact and activity locations beyond Estonia's national borders. At the same time, trips categorized as "visiting friends and relatives" are becoming more important in tourism (Yousuf & Backer, 2015).

One limitation of the current study is the lack of information about the quality of the contacts studied. When studying networks based on mobile phone–call activities, it must be recognized that the network does not reflect the individual's whole social network but only that part of it in which the person communicates via mobile phone. A person's social network may also include people (including important social relationships) with whom they do not communicate via mobile phone (e.g. neighbours, grandparents, etc.). In addition, people who do communicate via mobile phone sometimes may not know each other, and these contacts are thus not indicative of the quality of relationships. Mobile phone–based studies have previously evaluated the quality of relationships based on the number (Calabrese et al., 2011) and duration of calls (Onnela et al., 2011, Phithakkitnukoon et al., 2012). As such, mobile phone–based data do not reflect social networks but, rather, networks of calling partners, where changes in communication habits taking place in society cannot be neglected (Mascheroni, 2007; Line et al., 2011; Aguiléra et al., 2012; Kane et al., 2014).

To conclude, relationships between social networks and spatial mobility can be studied using different approaches, data sets and methods, and we can find a number of field-specific approaches (e.g. transportation studies, physics and sociology). The common theoretical and methodological framework, however, is still under development. There is a need for more research to gather existing and new knowledge about the dynamics and influence of social networks related to spatial mobility. This study confirms the importance of relationships between social networks and spatial mobility, and the influence of individual factors on the strength of these relationships. These findings confirm the capacity of mobile phone–based data sets to study and extend knowledge of the relationships between mobile communication networks and spatial mobility.

Acknowledgements

We are grateful for the collaboration in data collection and management to Positium LBS.

This chapter is a reprint of the journal article: Anniki Puura, Siiri Silm and Rein Ahas (2018) The Relationship between Social Networks and Spatial Mobility: A Mobile-Phone-Based Study in Estonia, *Journal of Urban Technology*, 25:2, 7–25, with the permission kindly granted by the publisher Taylor and Francis Ltd, www.tandfonline.com on behalf of The Society of Urban Technology.

Disclosure statement

No potential conflict of interest was reported by the authors

Funding

This research was funded by the institutional research grant [grant no. IUT 2–17] of the Ministry of Education and Research of Estonia; Research Professor grant of Estonian Academy of Sciences; grant Real-time Location-based Big Data Algorithms [grant no. 3.2.1201.13–0009] of Archimedes Foundation; the research grant "Urbanization, Mobilities and Immigration (URMI)" of Finnish Academy; and the Estonian Science Infrastructure Road Map project "Infotechnological Mobility Observatory (IMO)".

Bibliography

Aguiléra, A., Guillot, C., & Rallet, A. (2012). Mobile ICTs and physical mobility: Review and research agenda. *Transportation Research Part A: Policy and Practice, 46*(4), 664–672.

Ahas, R., Aasa, A., Roose, A., Mark, Ü., & Silm, S. (2008). Evaluating passive mobile positioning data for tourism surveys: An Estonian case study. *Tourism Management, 29*(3), 469–486.

Ahas, R., Silm, S., Järv, O., Saluveer, E., & Tiru, M. (2010). Using mobile positioning data to model locations meaningful to users of mobile phones. *Journal of Urban Technology, 17*(1), 3–27.

Borgatti, S. P., Mehra, A., Brass, D. J., & Labianca, G. (2009). Network analysis in the social sciences. *Science, 323*, 892–895.

Buliung, R. N., Roorda, M. J., & Remmel, T. K. (2008). Exploring spatial variety in patterns of activity-travel behaviour: Initial results from the Toronto Travel-Activity Panel Survey (TTAPS). *Transportation, 35*, 697–722.

Calabrese, F., Smoreda, Z., Blondel, V. D., & Ratti, C. (2011). Interplay between telecommunications and face-to-face interactions: A study using mobile phone data. *PLoS ONE, 6*(7), e20814.

Carrasco, J. A., Hogan, B., Wellman, B., & Miller, E. J. (2008). Agency in social activity interactions: The role of social networks in time and space. *Tijdschrift voor Economische en Sociale Geografie, 99*(5), 562–583.

Carrasco, J. A., & Miller, E. J. (2006). Exploring the propensity to perform social activities: A social network approach. *Transportation, 33*, 463–480.

Carrasco, J. A., Miller, E. J., & Wellman, B. (2008). How far and with whom do people socialize? Empirical evidence about distance between social network members. *Transportation Research Record: Journal of the Transportation Research Board, 2076*, 114–122.

Castells, M. (1996). *The rise of the network society: The Information age: Economy, society, and culture:* Vol. I. Oxford: Blackwell.

Castells, M. (2010). *The rise of the network society: The information age: Economy, society, and culture*: Vol. I, 2nd ed. with a New Preface. Oxford: Wiley-Blackwell.

Cooke, T. J. (2013). Internal migration in decline. *The Professional Geographer, 65*(4), 664–675.

Cristaldi, F. (2005). Commuting and gender in Italy: A methodological issue. *The Professional Geographer*, *57*, 268–284.

Dijst, M. (1999). Two-earner families and their action spaces: A case study of two Dutch communities. *GeoJournal*, *48*(3), 195–206.

Eurobarometer. (2013, February–March). *E-communications household survey, fieldwork*. Special Eurobarometer 396/Wave EB79.1 (TNS Opinion & Social). Retrieved from http://ec.europa.eu/digital-agenda/en/news/special-eurobarometer-396-e-communications-household-survey. (Accessed December 8, 2016).

European Parliament. (1995). Directive 95/46/EC of the European Parliament and of the Council of 24 October 1995 on the Protection of Individuals with regard to the processing of personal data and on the free movement of such data. *Official Journal L*, *281*, 31–50. Retrieved from http://eur-lex.europa.eu/LexUriServ/LexUriServ.do?uri=CELEX:31995L0046:en:HTML (Accessed December 8, 2016).

European Parliament. (2002). Directive 2002/58/EC of the European Parliament and of the Council of 12 July 2002 concerning the processing of personal data and the protection of privacy in the electronic communications sector (Directive on Privacy and Electronic Communications). *Official Journal L*, *201*, 37–47. Retrieved from http://eur-lex.europa.eu/LexUriServ/LexUriServ.do?uri=CELEX:32002L0058:en:HTML (Accessed December 8, 2016).

Eurostat. (2016a). *Statistics explained*. Retrieved from http://ec.europa.eu/eurostat/statistics-explained (Accessed December 8, 2016).

Eurostat. (2016b). *Gender pay gap statistics*. Retrieved from http://ec.europa.eu/eurostat/statistics-explained/index.php/Gender_pay_gap_statistics (Accessed December 8, 2016).

Expert, P., Evans, T. S., Blondel, V. D., & Lambiotte, R. (2011). Uncovering space-independent communities in spatial networks. *Proceedings of the National Academy of Sciences*, *108*(19), 7663–7668.

Flamm, M., & Kaufmann, V. (2006, March). The concept of personal network of usual places as a tool for analyzing human activity spaces: A quantitative exploration. Paper presented at 6th Swiss Transport Research Conference, March 15–17, Ascona.

Golledge, R. G., & Stimson, R. J. (1997). *Spatial behaviour: A geographic perspective*. New York, NY: Guilford Press.

Hägerstrand, T. (1970). What about people in regional science? *Papers of the Regional Science Association*, *24*(1), 6–21.

Hamilton, K. (2001, January). Gender and transport in developed countries. Paper presented at Gender Perspectives for Earth Summit 2002: Energy, Transport Information for Decision-Making, January 10–12, Berlin.

Heidemann, J., Klier, M., & Probst, F. (2012). Online social networks: A survey of a global phenomenon. *Computer Networks*, *56*(18), 3866–3878.

Järv, O., Ahas, R., & Witlox, F. (2014). Understanding monthly variability in human activity spaces: A twelve-month study using mobile phone call detail records. *Transportation Research Part C: Emerging Technologies*, *38*(1), 122–135.

Kamruzzaman, M., & Hine, J. (2012). Analysis of rural activity spaces and transport disadvantage using a multi-method approach. *Transport Policy*, *19*, 105–120.

Kane, G. C., Alavi, M., Labianca, G. J., & Borgatti, S. P. (2014). What's different about social media networks? A framework and research agenda. *MIS Quarterly*, *38*(1), 275–304.

Kowald, M., van den Berg, P. E. W., Frei, A., Carrasco, J. A., Arentze, T. A., Axhausen, K. W. . . . Wellman, B. (2013). Distance patterns of personal networks in four countries: A comparative study. *Journal of Transport Geography*, *31*, 236–248.

Kwan, M.-P., & Schwanen, T. (2016). Geographies of mobility. *Annals of the American Association of Geographers, 106*(2), 243–256.

Lambiotte, R., Blondel, V. D., de Kerchove, C., Huens, E., Prieur, C., Smoreda, Z., & van Dooren, P. (2008). Geographical dispersal of mobile communication networks. *Physica: A, 387*, 5317–5325.

Larsen, J., Urry, J., & Axhausen, K. W. (2006). *Mobilities, networks, geographies.* Aldershot: Ashgate Publishing Ltd.

Line, T., Jain, J., & Lyons, G. (2011). The role of ICTs in everyday mobile lives. *Journal of Transport Geography, 19*, 1490–1499.

Mascheroni, G. (2007). Global nomads' network and mobile sociality: Exploring new media uses on the move. *Information, Communication and Society, 10*(4), 527–546.

Masso, A., Silm, S., & Ahas, R. (2017, June). Spatial mobility in 'high-speed-societies': A study of generational differences with mobile phone data. Paper presented at the Swiss Mobility Conference, June 29–30, Lausanne.

Matous, P., Todo, Y., & Mojo, D. (2013). Boots are made for walking: Interactions across physical and social space in infrastructure-poor regions. *Journal of Transport Geography, 31*, 226–235.

McPherson, M., Smith-Lovin, L., & Cook, J. M. (2001). Birds of a feather: Homophily in social networks. *Annual Review of Sociology, 27*, 415–444.

Meurs, H., & Haaijer, R. (2001). Spatial structure and mobility. *Transportation Research Part D: Transport and Environment, 6*, 429–446.

Miranda-Moreno, L. F., Eluru, N., Lee-Gosselin, M., & Kreider, T. (2012). Impact of ICT access on personal activity space and greenhouse gas production: Evidence from Quebec City, Canada. *Transportation, 39*(5), 895–918.

Mok, D., Wellman, B., & Carrasco, J. A. (2010). Does distance matter in the age of the internet? *Urban Studies, 47*, 2747–2783.

Mokhtarian, P. L., Salomon, I., & Handy, S. L. (2006). The impacts of ICT on Leisure activities and travel: A conceptual exploration. *Transportation, 33*, 263–289.

OECD. (2016). *Glossary of statistical terms.* Retrieved from https://stats.oecd.org/glossary/index.htm (Accessed December 8, 2016).

Onnela, J.-P., Arbesman, S., Gonzalez, M. C., Barabasi, A.-L., & Christakis, N. A. (2011). Geographic constraints on social network groups. *PLoS ONE, 6*(4), e16939.

Palchykov, V., Kaski, K., Kertész, J., Barabasi, A.-L., & Dunbar, R. I. M. (2012). Sex differences in intimate relationships. *Scientific Reports, 2*(370). doi:10.1038/srep00370

Phithakkitnukoon, S., Smoreda, Z., & Olivier, P. (2012). Socio-geography of human mobility: A study using longitudinal mobile phone data. *PLoS ONE, 7*(6), e39253.

Polk, M. (1996, October). Swedish men and women's mobility patterns: Issues of social equity and ecological sustainability. In *Proceedings from the Second National Conference on Women's Travel Issues* (pp. 23–26), October, Baltimore.

Puura, A., Silm, S., & Ahas, R. (2018). The relationship between social networks and spatial mobility: A mobile-phone based study in Estonia. *Journal of Urban Technology, 25*(2), 7–25.

Ratti, C., Sobolevsky, S., Calabrese, F., Andris, C., Reades, J., Martino, M. . . . Strogatz, S. H. (2010). Redrawing the map of Great Britain from a network of human interactions. *PLoS ONE, 5*(12), e14248.

Saramäki, J., Leicht, E. A., Lopez, E., Roberts, S. G. B., Reed-Tsochas, F., & Dunbar, R. I. M. (2014). The persistence of social signatures in human communication. *Proceedings of the National Academy of Sciences of the United States of America, 111*, 942–947.

Scheiner, J., & Kasper, B. (2003). Lifestyles, choice of housing location and daily mobility: The lifestyle approach in the context of spatial mobility and planning. *International Social Science Journal, 55*(176), 319–332.

Schönfelder, S., & Axhausen, K. W. (2003). Activity spaces: Measures of social exclusion? *Transport Policy, 10*(4), 273–286.

Schönfelder, S., & Axhausen, K. W. (2010). *Urban rhythms and travel behaviour: Spatial and temporal phenomena of daily travel.* Aldershot: Ashgate Publishing Ltd.

Sharmeen, F., Arentze, T., & Timmermans, H. (2015). Predicting the evolution of social networks with life cycle events. *Transportation, 42,* 256–268.

Shi, L., Wu, L., Chi, G., & Liu, Y. (2016). Geographical impacts on social networks from perspectives of space and place: An empirical study using mobile phone data. *Journal of Geographical Systems, 18*(4), 359–376.

Silm, S., & Ahas, R. (2014). Ethnic differences in activity spaces: A study of out-of-home nonemployment activities with mobile phone data. *Annals of the Association of American Geographers, 104*(3), 542–559.

Silm, S., Ahas, R., & Nuga, M. (2013). Gender differences in space-time mobility patterns in a postcommunist city: A case study based on mobile positioning in the suburbs of Tallinn. *Environment and Planning B: Planning and Design, 40,* 814–828.

Takhteyev, Y., Gruzd, A., & Wellman, B. (2012). Geography of twitter networks. *Social Networks, 34,* 73–81.

Tillema, T., Dijst, M., & Schwanen, T. (2010). Face-to-face and electronic communications in maintaining social networks: The influence of geographical and relational distance of information content. *New Media & Society, 12*(6), 965–983.

Urry, J. (2003). Social networks, travel and talk. *British Journal of Sociology, 54*(2), 155–175.

van Acker, V., Wee, B., & Witlox, F. (2010). When transport geography meets social psychology: Toward a conceptual model of travel behaviour. *Transport Reviews, 30*(2), 219–240.

van den Berg, P., Arentze, T., & Timmermans, H. (2012). A multilevel path analysis of contact frequency between social network members. *Journal of Geographical Systems, 14,* 125–141.

van den Berg, P., Arentze, T., & Timmermans, H. (2013). A path analysis of social networks, telecommunication and social activity-travel patterns. *Transportation Research Part C: Emerging Technologies, 26,* 256–268.

Viry, G. (2012). Residential mobility and the spatial dispersion of personal networks: Effects on social support. *Social Networks, 34,* 59–72.

Wang, Y., Kang, C., Liu, Y., & Andris, C. (2015). Linked activity spaces: Embedding social networks in urban space. In M. Helbich, J. Jokar Arsanjani, & M. Leitner (Eds.), *Computational approaches for urban environments* (pp. 313–336). Switzerland: Springer International.

Wellman, B. (1988). Structural analysis: From method and metaphor to theory and substance. In S. Berkowitz & B. Wellman (Eds.), *Social structures: A network approach.* New York, NY: Cambridge University Press.

Wrzus, C., Hanel, M., Wagner, J., & Neyer, F. G. (2013). Social network changes and life events across the life span: A meta-analysis. *Psychological Bulletin, 139,* 53–80.

Yousuf, M., & Backer, E. (2015). A content analysis of Visiting Friends and Relatives (VFR) travel research. *Journal of Hospitality and Tourism Management, 25,* 1–10.

Yuan, Y., Raubal, M., & Liu, Y. (2012). Correlating mobile phone usage and travel behavior: A case study of Harbin, China. *Computers, Environment and Urban Systems, 36,* 118–130.

6 On the relative importance of social influence in transport-related choice behaviour

Empirical evidence from three stated-choice experiments

Bilin Han, Jinhee Kim, Soora Rasouli and Harry Timmermans

Introduction

The study of social networks and social influence has been the core business of sociology and anthropology since their very beginnings. Many studies have identified social networks and analysed their properties (e.g., French, Raven, & Cartwright, 1959; Burnkrant & Cousineau, 1975; Turner & Oakes, 1986; Friedkin & Johnsen, 1990; Turner, 1991; Cialdini & Goldstein, 2004). Social networks in their spatial settings constitute the daily context in which people shape their lives. These networks generate and sustain social norms and, in that sense, regulate daily behaviour, particularly in small localized communities. The maintenance of social networks implies social interaction, which leads to an exchange of information and opinions and hence triggers the formation of common attitudes and opinions. In turn, attitudes and social norms regulate the dynamics of social networks.

Beyond sociology and anthropology, social influence has been a topic of interest in more applied sciences such as marketing and consumer sciences (e.g. Ibarra & Andrews, 1993; Venkatesh & Morris, 2000; Subramani & Rajagopalan, 2003; Dholakia, Bagozzi, & Pearo, 2004; Algesheimer, Dholakia, & Herrmann, 2005; Hudson, Huang, Roth, & Madden, 2016; Schivinski & Dabrowski, 2016; Hambrick & Lovelace, 2018). It has been realized that word of mouth, lifestyle segmentation, peer influence and similar processes influence purchasing behaviour, and hence, the role of social influence in purchasing behaviour has been widely examined.

Compared to these fields of study, transportation and travel behaviour research has been relatively late in examining social networks and social influence. It was not until a few years ago that transportation researchers started to investigate the relationship between social networks and activity-travel behaviour. One of the main reasons is that transportation witnessed a shift in paradigm in the late 1990s (Rasouli & Timmermans, 2014). It was realized that travel is directly derived from people's needs and desires to become involved in activity participation. Social activities were commonly considered as an important activity category. It was felt that social travel might be better predicted if social networks can be identified in terms of network size and the spatial distribution of network members. To maintain their social networks, individuals need to stay in contact. Dependent on distance/travel time, some contacts will involve physical travel.

Recently, a substantial number of articles that analysed different facets of social networks and associated travel behaviour has been published (for an overview, see Kim, Rasouli, & Timmermans, 2018). Similar to marketing research, other studies have explored the effects of social influence on different types of travel-related behaviour. Examples relate to travel-mode choice (Dugunji & Walker, 2005; Goetzke, 2008; Pike & Lubell, 2016), car ownership (Goetzke & Weinberger, 2012), the intention to purchase electric vehicles (Axsen & Kurani, 2012; Rasouli & Timmermans, 2013, 2016), attitudes and decisions about bicycling (Gordon & Handy, 2012; Sherwin, Chatterjee, & Jain, 2014), bicycle ownership (Maness & Cirillo, 2016) and the intention to joining car sharing (Kim, Rasouli, & Timmermans, 2017). It raises the question about the relative importance of social influence in travel-related decision-making and choice behaviour.

In this chapter, we give an answer to this question based on three studies that we completed over the last couple years. The first study is on job application decisions (Han & Timmermans, 2015). Insight into such decisions is relevant because job locations and homes constitute the anchors of daily action spaces. The focus of this study is on households with children. We are interested in the question to what extent opinions of social network members about the importance of work and about raising children influence the decision to apply for particular job profiles. The second study concerned people's intention to purchase an electric car (Rasouli & Timmermans, 2013, 2016; Kim, Rasouli, & Timmermans, 2014). The study investigated social influence associated with reviewers' opinion and the market shares of electric cars among different types of social networks, such as family, friends, co-workers and peers. The third study examines the effects of social influence in the decision to join a car-sharing organization (Kim et al., 2017) The study employed an egocentric network approach to identify social network members (i.e. alters) and used their information in a stated-choice experiment to measure the effect of alters' behaviour on the ego's decision.

These three studies on social influence are summarized in the remainder of this chapter. In each chapter, we state the aims and objectives of the study, outline the data collection, describe some key properties of the sample and then discuss the main results of the study with special focus on social influence. Several different analyses of varying complexity have been applied to these data sets. In this chapter, we report the findings of the most advanced analysis. Readers are referred to the original publications for the details of complementary analyses. Finally, we complete this chapter by drawing some overall conclusions about the relevance importance of social influence in travel-related choices and discuss implications for future research.

Objectives of the three studies

Study 1: job application decisions

The aim of this study was to estimate the effects of job attributes, social influence and socio-demographic characteristics on job application decisions. To that effect, a stated-choice experiment, which systematically varied different job attributes and opinions of different members of one's social network, was developed. A mixed binary logit model was estimated to estimate the relative effects of social influence on the probability to apply for the job.

Study 2: intention to buy an electric car

The second study concerned people's intention to purchase an electronic car. Similar to for example brand-names items, we assumed that the popularity of electric cars might be influenced by the popularity of the electric car among social network members. Thus, the aim of this study was to predict the intention to buy an electric car as a function of (1) its attributes; (2) context; (3) the reviews it receives; (4) acceptance of the electric car by relatives, friends, co-workers and peers; and (5) socio-demographic variables. As in the first study, a stated-choice experiment was designed to collect the necessary data and derive the relative importance of these factors. Considering the objective of studying the impact of social influence, the design of the experiment differed from conventional stated-choice designs, which typically vary only the attributes of the choice alternatives.

Study 3: decision to join a car-sharing organization

The objective underlying the third study was to investigate social influence in car-sharing decisions. Using an egocentric approach, social relationships of respondents were identified, and the social distance for each ego–alter relationship was established. A sequential stated adaptation experiment was created to explicitly capture social influence in terms of a change in choice behaviour between consecutive tasks. The estimated social distance and the alters' choices were incorporated into n social influence variable of the car-sharing decision model.

Comparison of the data used

The data for Study 1 were collected in January 2015 in the Netherlands among a sample of 1575 respondents. Dual-earner households with young children were oversampled because this category was deemed most important in light of the objectives of the study. Ultimately, after cleaning the data and selecting relevant households, 1,051 respondents were used for the final analysis. The data for Study 2 were collected in June 2012 in the Netherlands. It involved a random sample of 726 respondents. Last, the data for Study 3 were collected in April 2015 in the Netherlands. The data of 955 target respondents, who have both a driver license and at least one car in their household, were used.

Table 6.1 shows the comparison of the sample characteristics of the three studies. We observed some differences in the selected individual and household characteristics, which are due to the sample selection criteria that somewhat differed, due to the differences in the objectives of the studies. For instance, since Study 1 specifically focused on two-adult households with children younger than 12 years, all households are composed of parents and at least one child, whilst the percentage of these households is only 39.9% in Study 2 and 27.7% in Study 3, respectively. In addition, the percentages of presence of children younger than 12 years are even less, 17.6% and 15.4% in Study 2 and Study 3, respectively. For the same reason, the age distribution also differs among the three studies.

Comparison of the choice experiments

All three studies were based on stated-choice experiments that systematically varied a set of attributes of the choice options and social influence attributes across different hypothetical choice situations. Stated preference and choice methods rely on the use of experimental designs to measure individual preferences or choices (e.g. Timmermans, 1984; Louviere, 1988; Louviere & Timmermans, 1990a, 1990b). These methods have been found particularly useful when historical data of the choice behaviour of interest or missing or when the researcher has reason to control the position of the data point in parameter space. The challenge is to design an experiment that allows estimating the model of interest and is valid and easy to complete. Orthogonal fractional factorial designs vary the attribute levels of a subset of attribute profiles such that all inter-attribute correlations are zero. In the case of measuring choices, the attribute profiles need to be put into choice sets such that the assumed choice model can be estimated. For a long time, these orthogonal designs have dominated applications. More recently, D-optimal designs have become more popular. These designs minimize the generalized variance of the parameter estimates of a pre-specified choice model by maximizing the determinant of the information matrix.

Table 6.2 gives an overview of the experimental designs used in the three studies. Study 1 considered two alternatives: whether to apply for a job described by five attributes, including the number of work hours/week, flexibility/work at home, interest in the job, salary and travel distance from home to work. Study 2 also considered two alternatives regarding the purchase of an electric car: purchase or not. The following attributes were systematically varied in the experiment to describe the car: the net capital price of the electric car, net operating cost, cruising range of the car, the time required to (re)charge the battery, the top speed of the car and the distance to a charging station. Study 3 designed the experiment with three alternatives, including do nothing, buy a second car and join a car-sharing organization. The selected attributes include purchase, maintenance and operating costs for each alternative, access time and availability of car use, which represents the possible non-availability of shared cars.

Having varied these attributes for describing each alternative, the three studies designed social influence attributes to represent opinions/behaviour of social network members that are assumed to influence individual decision-making.

Table 6.1 Sample characteristics

Category	Attribute	Levels	Study 1: job application (N=1051)	Study 2: electric car (N=726)	Study 3: car-sharing (N=955)
Individual characteristics	Gender	Male	43.7%	48.8%	47.3%
		Female	56.3%	51.2%	52.7%
	Age	18–25 years	4.7%	12.5%	10.5%
		26–35 years	40.7%	16.8%	16.2%
		>35 years	54.6%	70.6%	73.3%
			(36–55 years: 53.6%; >55 years: 1.0%)	(36–50 years: 31.0%; >51 years: 39.6%)	(36–50 years: 23.5%; >51 years: 49.8%)
	Education level	Elementary school	0.4%	1.5%	1.8%
		Lower vocational school	5.5%	9.2%	6.2%
		Middle general education	8.3%	14.4%	15.2%
		Middle specialized education	7.2%	11.0%	11.1%
		Middle vocational education	32.7%	26.6%	27.1%
		Higher vocational education	35.0%	26.2%	29.9%
		University	10.9%	10.6%	8.4%
		Others	0.0%	0.5%	0.3%
Household characteristics	Number of household members	1 person	0.0%	19.6%	17.1%
		2 persons	0.0%	36.0%	46.1%
		3 persons	34.8%	14.7%	15.9%
		More than 3 persons	65.2%	29.8%	20.9%
	Composition	Single without children	0.0%	28.5%	17.1%
		Single with children	0.0%		4.3%
		Couple without children	0.0%	31.5%	42.0%
		Couple with children	100%	39.9%	27.7%
		Others	0.0%	0.0%	8.9%
	Presence of children <= 12 years	Yes	100.0%	17.6%	15.4%
		No	0.0%	82.4%	84.6%

Table 6.2 Design of stated-choice experiment

Category		Study 1: job application	Study 2: electric car	Study 3: car-sharing
Alternatives		1: Yes (apply for the job) 2: No	1: Yes (purchase the electric car) 2: No	1: Do-nothing 2: Buy 2nd car 3: Join car-sharing
Number of attributes & their levels	For describing alternatives	5 attributes with 4 levels	2 attributes with 8 levels; 4 attributes with 4 levels	13 attributes with 4 levels
	For representing social influences	4 attributes with 4 levels	5 attributes with 4 levels	3 attributes with 4 levels
Experiment design		Orthogonal design	Orthogonal design	D-efficient design
Number of choice situations per a respondent		8	16	

Table 6.3 provides the designed social influence attributes and their levels for each study. In Study 1, social influence attributes were defined as opinions/attitudes of members of the respondent's social network, differentiated into parents, relatives, friends, and peers. These opinions concerned the time allocation of parents between work and children-related activities and opinions about gender roles. In Study 2, the social influence attributes describe possible reviews and adoption of this new technology by various members of social networks (family, friends, co-workers and the larger social network of peers) and the impact of the nature of reviews (positive or negative). In Study 3, social influence was represented in terms of the number of social network members who chose to join a car-sharing organization in the same choice situation. During the experiment, actual names of a respondent's social network members who chose to join a car-sharing organization were shown to the respondent. The actual names were collected via an egocentric-network data-collection approach, and the choices of social networks were hypothetically designed and varied across choice situations.

Comparison of the estimated results

Table 6.4 provides an overview of the model specifications and goodness-of-fit of the three studies. Study 1 employed a mixed logit model to allow for heterogeneity in people's taste regarding working conditions (e.g., work hours, flexibility, salary, etc.). In addition, 12 non-random parameters were estimated to measure the effects of the opinions of social network members on the job application decision. Study 2 estimated a mixed logit model with 16 random parameters for social influence. The study examined heterogeneity in social influence that varies across individuals. Study 3 also considered heterogeneity in social influence. The study

Table 6.3 Experimental attributes for representing social influence

Study	Attribute	Attribute levels
Study 1: job application	Social network members' opinions about work and child care: (1) Parents' opinion (2) Relatives' opinion (3) Friends' opinion (4) Peers' opinion	Level 1: full-time work Level 2: spending more time on work than for child care Level 3: spending more time on child care than work Level 4: giving priority to child care
Study 2: electric car	Reviewers' opinion about electric vehicles	Level 1: Only positive Level 2: Mainly positive, but some criticism Level 3: Mainly negative, but some positive Level 4: Only negative
	Share of electric vehicles among: (1) Friends (2) Relatives (3) Co-workers (4) Peers	Level 1: 0% Level 2: 25% Level 3: 50% Level 4: 75%
Study 3: car-sharing	The number of social network members who chose to join car-sharing in the same choice situation: (1) Family/relatives (living separately) (2) Friends (3) Others (co-workers, neighbours)	Level 1: Nobody Level 2: 1 person (with actual name) Level 3: 2 persons (with actual names) Level 4: 3 persons (with actual names)

Table 6.4 Model specification and goodness of fit

Category		Study 1: job application	Study 2: electric car	Study 3: car-sharing
Type of model		Mixed logit	Mixed logit	Hybrid choice
Number of parameters in the choice model	Alternative-attribute	16 (including 16 random parameters)	27 (including 1 random parameter)	11 (including 2 random parameters)
	Social influences	12	16 (including 16 random parameters)	3
	Socio-demographics	26	20	18
Goodness-of-fit (ρ^{-2})		0.310	0.349	0.627

focused on heterogeneity in social distance representing the strength of the relationship between individuals. Social distance was regarded as a latent variable that is not directly observed and involves a probability density function. A hybrid choice model was developed to identify simultaneously the latent social distance and the social influence on the car-sharing decision. As shown in Table 6.4, the goodness-of-fit values in the three studies are relatively high, implying that the proposed models have high explanatory power.

In this section, we discuss the estimation results of the three studies, focusing mainly on social influence parameters. Table 6.5 summarizes the estimation results of the studies with respect to the effects of social influence. The estimation results of Study 1 show that none of the estimated effects of social influence are significant at the 5% significance level. Yet, the probability of applying for a job profile decreases when parents and relatives stimulate taking a full-time job or spending more time on work than on child care, while the utility for a job profile increases if they stimulate spending more time on child care or give priority to child care. As for the social influence of friends, the utility of a job profile decreases when friends stimulate spending more time on work or on child care. In the case of peers, the effects are positive, except when they stimulate spending more time on child care.

The estimation results of Study 2 show that the estimated part-worth utility of reviews monotonically decreases from more positive to more negative reviews. The function is almost linear, except for "only negative reviews". The estimated part-worth utilities for the different sources of social influence differ, both in curvature and absolute size. The influence seems highest for friends and lowest for co-workers. Estimated coefficients first increase with increasing market shares of electric cars among friends and then substantially drop for the 75% category. In the case of relatives, the estimated utility increases with increasing market share to satiate at 50%. The curvature for co-workers is first monotonically decreasing with increasing market share and then increases again when the market share is high. Finally, the curve for peers shows that estimated part-worth utilities decrease with increasing market shares of electric cars among peers and then slightly increases again.

Furthermore, the estimation results suggest heterogeneity in social influence. In particular, scale parameters vary widely for the effects of different members of the social network. In the case of friends, heterogeneity is highest when the market share of the electric car is 25%. The scale parameter shows a less extreme distribution in the case of relatives. It first increases from zero market share to 25% market share and then drops when the market share of the electric car among relatives is 50%. Heterogeneity in the effect of co-workers shows more variation when the share is 0% and less variation when the shares are, respectively, 25% and 50%. Finally, the scale parameter monotonically increases in the case of increasing market share of the electric car among peers, which is the reverse of the co-workers pattern. Most random effects are significant.

The estimation results of Study 3 reveal that the parameters for family and friends are significant and positive, indicating that the decision to join a car-sharing organization is positively influenced by the choices of family and friends. In other

Table 6.5 Estimation results for the effect of social influence

Study 1: Estimated social influence in job applications

Attributes		Estimates	p-value
Parents' opinion	Full-time work	−0.0791	0.3768
	Spending more time on work than for child care	−0.0165	0.8550
	Spending more time on child care than work	0.0475	0.5970
	Giving priority to child care	0.0481	−
Relatives' opinion	Full-time work	−0.0079	0.9293
	Spending more time on work than for child care	−0.0708	0.4265
	Spending more time on child care than work	0.0604	0.5036
	Giving priority to child care	0.0183	−
Friends' opinion	Full-time work	−0.0041	0.9632
	Spending more time on work than for child care	0.0237	0.7958
	Spending more time on child care than work	0.0147	0.8724
	Giving priority to child care	−0.0343	−
Peers' opinion	Full-time work	0.0565	0.5344
	Spending more time on work than for child care	0.0179	0.8439
	Spending more time on child care than work	−0.0820	0.3561
	Giving priority to child care	0.0076	−

Study 2: Estimated social influence in the purchase of EVs

Attributes		Estimates	P-value
Reviewers' opinion	Only positive reviews	M: **1.0173**	M: 0.0000
		S: **2.0717**	S: 0.0000
	Mainly positive reviews but also some criticism	M: **0.5380**	M: 0.0000
		S: **2.5394**	S: 0.0000
	Mainly negative reviews but also positive	M: −0.1266	M: 0.1291
		S: **2.8141**	S: 0.0000
	Only negative reviews	M: −1.4287	−
Share of EVs among friends	0%	M: 0.1047	M: 0.2051
		S: **0.9345**	S: 0.0000
	25%	M: 0.1881	M: 0.0222
		S: **4.0219**	S: 0.0000
	50%	M: **0.5365**	M: 0.0000
		S: **0.4502**	S: 0.0000
	75%	M: −0.8293	−
Share of EVs among relatives	0%	M: **−0.3063**	M: 0.0002
		S: **1.2897**	S: 0.0000
	25%	M: −0.0609	M: 0.4767
		S: **1.4456**	S: 0.0000
	50%	M: 0.1847	M: 0.0239
		S: **0.5762**	S: 0.0000
	75%	M: 0.1825	−

Study 2: Estimated social influence in the purchase of EVs

Attributes			Estimates	P-value
Share of EVs among co-workers	0%		M: 0.0665	M: 0.4492
			S: **1.6388**	S: 0.0000
	25%		M: 0.0015	M: 0.9846
			S: **0.7983**	S: 0.0000
	50%		M: −0.2055	M: 0.0157
			S: **0.4060**	S: 0.0000
	75%		M: 0.1375	–
Share of EVs among peers	0%		M: **0.3203**	M: 0.0001
			S: **0.7934**	S: 0.0000
	25%		M: 0.1009	M: 0.2358
			S: **0.9640**	S: 0.0000
	50%		M: **−0.2640**	M: 0.0013
			S: **2.8347**	S: 0.0000
	75%		M: −0.1572	–

Study 3: Estimated social influence in joining a car-sharing organization

Attributes		Estimates	P-value
The number of social network members who joined car sharing	Family/relatives (living separately)	**0.0607**	0.0000
	Friends	**0.0500**	0.0000
	Others (co-workers, neighbours)	0.0005	0.9638

Note: Estimates whose p-values are less than 0.01 are marked in bold. "M" and "S" indicate the estimated mean value and scale parameter of a random parameter, respectively. EV = electric vehicle.

words, people tend to be more willing to join a car-sharing organization when more family members and friends joined before. On the other hand, the choices of other social categories (co-workers, neighbours, etc.) do not significantly affect individual choice behaviour.

Conclusion and discussion

The study of social influence has been of paramount importance in marketing and consumer studies and, of course, in sociology and anthropology. Relatively late, transportation researchers have started to show some interest in social networks and social influence. The recent interest in this phenomenon in transportation research raises the question of whether this concept is of equal importance in explaining transport-related choices and decision-making processes than in other application domains.

In this chapter, we discuss selected evidence of three of our studies on social influence. The first study pertains to job application decisions; the second, to the intention to buy an electric car; and the third, to the decision of joining a car-sharing organization. Detailed results of these studies have been published before.

In this chapter, we summarize the results and concentrate on the comparison of the studies.

Results of all three studies evidence that the relative importance of social influence variables is low compared to the importance of the attributes of the choice alternatives, in general, and compared to costs considerations, in particular. Yet, we observe some differences. The effect of social influence in the car-sharing decision tends to be higher (more substantial) than in the context of the purchase of an electric car and the job application decision. In turn, the electric car decision tends to be more influenced by the social network and opinions of social network members than the job application decision. This may indicate that the strength of social influence is associated with the familiarity and popularity of the decision problem. In other words, people tend to be more influenced by others when the choice alternatives are unfamiliar to them. Car sharing is less familiar than electric vehicles, and people are less familiar with purchasing electric vehicles than job applications.

The results also indicate that the effect of social influence is related to the importance of the decision to individuals. Job applications have long-term repercussions for individuals, and therefore, ultimately, personal preferences are dominant in the final decision. Purchasing an electric car is a more expensive investment than is joining a car-sharing organization.

Thus, although there is evidence of social influence, the effects of social influence are less strong than reported in many marketing contexts. One may only speculate why social influence plays a less prominent role in travel-choice contexts. First, the classic marketing examples mostly relate to fashion brands and electronics targeted at teenagers and young adults. It is the time when they try to find their identity either by differentiating themselves in lifestyle and supporting attitudes from their older generations or from other same-cohort groups, in part reflected in clothes and other consumer products. In some sense, they lose their identity by adopting the behaviour of their lifestyle segment and social group. The choice behaviour that we examined in our studies and the age groups are quite remote from these examples. We are looking across age groups, and the object of choice lacks the features that express a clear lifestyle group.

Second, although often relatively expensive, the merchandise involved in strong social branding includes low involvement goods that are bought regularly. In contrast, the purchase of a car is a high-involvement decision. People regret wrong choices, and the financial burden is high. In addition, we know from the literature that many people are still concerned about the new technology, the higher price, low battery life, the lower range and the lack of charging stations. Consequently, social influence plays a lesser role. Similarly, one's job and mobility tools touch at the very heart of people's daily life. If social influence plays a role, it is only marginal and comes into effect only when the key requirements are met.

It should be emphasized that all our studies are based on stated-choice experiments. While such experiments have the potential advantage that researchers can control the distribution of data points in parameter space, results depend on the realism of representing social influence. If our findings would be generalizable, what are the implications of the relatively small effects of social influence for future studies on social influence in a transportation context? First, we should emphasize that the number of studies on social influence in transportation research is still quite limited. Hence, it is important to replicate our findings in the same and other choice context, perhaps further improving data collection and the measurement of social influence. Second, even if the effect is small, it is important to account for it, particularly when word of mouth and diffusion through social networks dominate the dynamics in market shares. Thus, transportation researchers should continue their efforts to quantify and qualify the effects of social influence on activity-travel choice behaviour.

References

Algesheimer, R., Dholakia, U. M., & Herrmann, A. (2005). The social influence of brand community: Evidence from European car clubs. *Journal of Marketing, 69*(3), 19–34.

Axsen, J., & Kurani, K. S. (2012). Interpersonal influence within car buyers' social networks: Applying five perspectives to plug-in hybrid vehicle drivers. *Environment and Planning A, 44*(5), 1047–1065.

Burnkrant, R. E., & Cousineau, A. (1975). Information and normative social influence in buyer behavior. *Journal of Consumer Research, 2*(2), 206–215.

Cialdini, R. B., & Goldstein, N. J. (2004). Social influence: Compliance and conformity. *Annual Review of Psychology, 55*, 591–621.

Dholakia, U. M., Bagozzi, R., & Pearo, L. K. (2004). A social influence model of consumer participation in network- and small-group-based virtual communities. *International Journal of Research in Marketing, 21*(3), 241–263.

Dugunji, E. R., & Walker, J. (2005). Discrete choice with social and spatial network interdependencies: An empirical example using mixed generalized extreme value models with field and panel effects. *Transportation Research Record: Journal of the Transportation Research Board, 1921*, 70–78.

French J. R. P., Raven, B., & Cartwright, D. (1959). The bases of social power. *Studies in Social Power*, 150–167.

Friedkin, N. E., & Johnsen, E. C. (1990). Social influence and opinions. *The Journal of Mathematical Sociology, 15*(3–4), 193–206.

Goetzke, F. (2008). Network effects in public transit use: Evidence from a spatially autoregressive mode choice model for New York. *Urban Studies, 45*(2), 407–417.

Goetzke, F., & Weinberger, R. (2012). Separating contextual from endogenous effects in automobile ownership models. *Environment and Planning A, 44*(5), 1032–1046.

Gordon, E., & Handy, S. L. (2012). Safe and normal: Social influences on the formation of attitudes toward bicycling. *Proceedings of the 91st Annual Meeting of the Transportation Research Board*, Washington, DC (CD-ROM).

Hambrick, D. C., & Lovelace, J. B. (2018). The role of executive symbolism in advancing new strategic themes in organizations: A social influence perspective. *Academic Management Review, 43*(1), 110–131.

Han, B., & Timmermans, H. J. P. (2015). Modeling two-adults household with children job application decisions under social influence. Paper presented at the HKSTS Meeting, Hong Kong.

Hudson, S., Huang, L., Roth, M. S., & Madden, T. J. (2016). The influence of social media interactions on consumer-brand relationships: A three-country study of brand perceptions and marketing behaviors. *International Journal of Research in Marketing, 33*(1), 27–41.

Ibarra, H., & Andrews, S. B. (1993). Power, social influence, and sense making: Effects of network centrality and proximity on employee perceptions. *Administrative Science Quarterly, 38*(2), 277–303.

Kim, J., Rasouli, S., & Timmermans, H. J. P. (2014). *Transportation Research Part A, 69,* 71–85.

Kim, J., Rasouli, S., & Timmermans, H. J. P. (2017). Investigating heterogeneity in social influence by social distance in car-sharing decisions under uncertainty: A regret-minimizing hybrid choice model framework based on sequential stated adaptation experiments. *Transportation Research Part C: Emerging Technologies, 85,* 47–63.

Kim, J., Rasouli, S., & Timmermans, H. J. P. (2018). Social networks, social influence and activity-travel behavior: A review of models and empirical evidence. *Transport Reviews, 38*(4), 499–523.

Louviere, J. J. (1988). *Analyzing decision making: Metric conjoint analysis.* Newbury Park, CA: Sage Publications.

Louviere, J. J., & Timmermans, H. J. P. (1990a). A review of recent advances in decompositional preference and choice models. *Tijdschrift voor Economische en Sociale Geografie, 81*(3), 214–225.

Louviere, J. J., & Timmermans, H. J. P. (1990b). Stated preference and choice models applied to recreation research: A review. *Leisure Sciences, 12*(1), 9–32.

Maness, M., & Cirillo, C. (2016). An indirect latent informational conformity social influence choice model: Formulation and case study. *Transportation Research Part B, 93,* 75–101.

Pike, S., & Lubell, M. (2016). Geography and social networks in transportation mode choice. *Journal of Transport Geography, 57,* 184–193.

Rasouli, S., & Timmermans, H. J. P. (2013). The effect of social adoption on the intention to purchase electric cars: A stated choice approach. *Transportation Research Record: Journal of the Transportation Research Board, 2344,* 10–19.

Rasouli, S., & Timmermans, H. J. P. (2014). Activity-based models of travel demand: Promises, progress and prospects. *International Journal of Urban Sciences, 18*(1), 31–60.

Rasouli, S., & Timmermans, H. J. P. (2016). Influence of social networks on latent choice of electric cars: A mixed logit specification using experimental design data. *Networks and Spatial Economics, 16*(1), 99–130.

Schivinski, B., & Dabrowski, D. (2016). The effect of social media communication on consumer perceptions of brands. *Journal of Marketing Communications, 22*(2), 189–214.

Sherwin, H., Chatterjee, K., & Jain, J. (2014). An exploration of the importance of social influence in the decision to start bicycling in England. *Transportation Research Part A, 68,* 32–45.

Subramani, M. R., & Rajagopalan, B. (2003). Knowledge-sharing and influence in online social networks via viral marketing. *Communications of the ACM, 46*(12), 300–307.

Timmermans, H. J. P. (1984). Decompositional multiattribute preference models in spatial choice analysis: A review of some recent developments. *Progress in Human Geography, 8*(2), 189-221.

Turner, J. C. (1991). *Social influence.* Milton Keynes, UK: Open University Press.

Turner, J. C., & Oakes, P. J. (1986). The significance of the social identity concept for social psychology with reference to individualism, interactionism and social influence. *British Journal of Social Psychology, 25*(3), 237–252.

Venkatesh, V., & Morris, M. G. (2000). Why don't men ever stop to ask for directions? Gender, social influence, and their role in technology acceptance and usage behavior. *MIS Quarterly, 24*(1), 115–139.

7 Social networks in transit

Exploring the development of new network-based travel practices

Tom Erik Julsrud and Matthew Hanchard

Mobile technologies social networks and travel activities

Since the early 1990s, the integration of the Internet and mobile media technologies has transformed much of the way individuals communicate with friends and relatives and their use of travel time. To clarify, mobile communication grew rapidly in most Western countries during the latter part of the 1990s, based mostly on regular telephone calls and text messaging (Castells, Fernandéz-Ardévol, Qiu, & Sey, 2007; Fernandez & Usero, 2009). While traditional telephony was typically carried out as a place-to-place communication episode, the mobile telephone represented a shift toward a new personal communication device which an individual could carry about. Although regular phone conversation remained the dominant form of communication throughout the 1990s, SMS text messaging became particularly popular amongst public transport travellers since it afforded silent and (almost-)real-time communication whilst on the move (Igarashi, Takai, & Yoshida, 2005; Julsrud & Bakke, 2008; Kim, Kim, Park, & Rice, 2007).

The introduction of smartphones at the turn of the century opened the way for a new type of mobile communication, including all kinds of Internet-based service applications (apps). While smartphones support traditional telephony and the SMS text-messaging of mobile phones, they also provide access (often via apps) to news channels, websites, blogs, films, social networking platforms and games. Throughout the 2000s and 2010s, there have been several advances in smart technology. For example, smart tablets now offer an alternative format of personal and mobile media device, while wearable smart technologies offer increased personalization and ubiquity (Billinghurst, Reichherzer, & Nassini, 2016; Li, Liu, Liu, Chen, & Ma, 2016). Alongside a growing diversity of media devices, the affordances they offer have grown too. Smart devices now offer the experience of augmented reality (Jurgenson, 2012, pp. 86–88), connect users to an emerging set of smart city infrastructures or 'Internet of Things' (DaCosta, 2013) and enable a quantification of the body for self-monitoring (Lupton, 2016). However, for now at least, the smartphone remains the most popular personal communication device or when travelling (Gössling, 2018).

Alongside the technical development of personal media devices, the Internet has expanded in scope and scale too, with Internet-based communication emerging as a central mode of interaction. In the context of travel for example, a recent study by Julsrud and Denstadli (2017) provides evidence that 40% of public transport users engage with social media networks while travelling on public transport in Oslo (see Figure 7.1), with 75% using SMS text messaging.

The diffusion of personal media has stoked debate amongst social scientists over their impact on the actions of local communities, family units and wider sets of social relations. A key them has been how communication media may make travel redundant causing, with potential risks and benefits. Early observers warned that Internet-based communication might replace face-to-face interaction with negative consequences for social engagement (Cairncross, 1997; Kraut, 1998; Turkle, 1995). From a sustainability perspective a substitution of travels by use of communication media was expected to the reduce number of trips and help to reduce traffic congestions and emissions (Mokhtarian, 2003; Mokhtarian & Salomon, 2001; Mokhtarian & Varma, 1998). However, there is little evidence to support the views that mediated communication is replacing face-to-face interactions or meetings.

More recent empirical studies suggested mobile media might lead to a new "technoscape" where existing relationships could be supported by an additional layer of mediated communication and a "connected presence" (Baym, 2010; Boase, 2008; Licoppe & Smoreda, 2004). For instance, Boase (2008) argued for a direct correlation where individuals who communicate more frequently face-to-face also communicated more frequently via virtual media. In contrast to the hypothesis that communication technology may substitute for travels, an increased

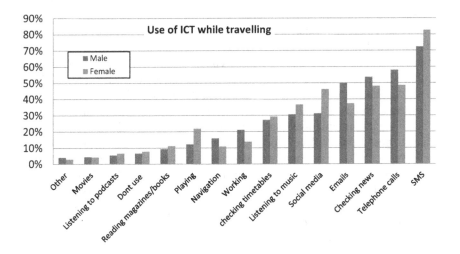

Figure 7.1 Mobile media and social media use among public transport users in Oslo and Trondheim

Source: Based on Julsrud and Denstadli (2017).

uptake and use of media suggest that this generate more, rather than fewer, trips to support distant relations (Choo & Mokhtarian, 2005).

The last years, attention has turned away from the possibilities for communication technologies to replace or generate travels to how it may alter travelling as a social practice and on how people interact in modern networked societies. On one hand, the question is raised as to how social networks and relations are transformed as people get more mobile lifestyles (Chen & Wellman, 2010; Kim, Rasouli, & Timmermans, 2018; Viry, 2012). In this area studies have investigated how digital communication, in combination with transport technologies and infrastructure, have influenced on network structures, activity patterns and settlement structures. In particular, the idea that people with a high degree of mobility are developing more fragmented and dispersed networks has been examined and been supported by empirical works (Ohnmacht, Götz, Schad, Haefeli, & Stettler, 2008; Viry, 2012). Another issue is how the emergence of personal media seems to alter the rhythms and regularity of the face-to-face communication. Studies of household consumption patterns have found that personal media use plays a significant role in the coordination of everyday activities, contributing to new ways of managing interpersonal relationships within family units (Christensen, 2009; Jensen, Sheller, & Wind, 2015; Licoppe & Smoreda, 2004). A similar transformation in interaction frequencies and interactions has been found among professionals (Larsen, Urry, & Axhausen, 2008; Schönfelder & Axhausen, 2010).

On the other hand, research has also addressed the question of how mobile technologies influence the travel activities and practices as such. Key questions here have been on how personal media use affects individuals' use of time when travelling, how travel is organized and practised through these media and its impact on the meaning and status of travel itself. Recent studies argue that transport carriers (bus stations, train stations, airports) have become "places" where diverse activities are carried out and which provide opportunities for new forms of multitasking with varying degrees of complexity (Kenyon & Lyons, 2007). Empirical investigations have found that access to mobile communication technologies (personal media) informs passengers' use of time as well as travel satisfaction (Axtell, Hislop, & Whittaker, 2008; Gripsrud & Hjorthol, 2006; Julsrud & Denstadli, 2017; Vilhelmson, Thulin, & Fahlén, 2011). Related studies have focused more closely on activity patterns and travelling over time and found that activities once closely related to geographical place (work, education and leisure) are becoming increasingly fragmented into multiple, smaller time slots and spread across many different places (Alexander, Ettema, & Dijst, 2010; Lenz & Nobis, 2007). In this, personal media are seen to spur a shift toward more frequent 'micro-coordination' and spontaneous patterns of mobility (Ling & Yttri, 2004; Rheingold, 2002).

This chapter takes hold in the recent streams of "micro-oriented" studies, zooming in on media use and its influence on peoples' travel activities. Our study follows up on the idea that mobile communication technology transforms the way travel is conducted. However, rather than testing any of the previously mentioned theories, we explore further the connections among travelling, the use of communication tools and how this is related to development and sustainment of social relations. The rationale for this more exploratory approach is, first, that the fast uptake and use of new technology (in particular, mobile applications) are triggering new routines and habits that may be more dominant ahead. An inductive approach is usually seen as more appropriate to grasp such "micro-trends" and emerging changes. Second, there are currently a few studies that address the connection between the structural and the travel-oriented approaches, mentioned earlier. By addressing people's social networks, as well as their everyday travel behaviour, we seek to bring studies of social network changes closer to the discussions of emerging network-based travels. Thus, we draw on more overarching theories of social structuration, network theories and theories of social practice and social interaction chains.

More explicitly, this chapter raises three interrelated research questions: (1) How does personal and social media use influence the way social relationships are sustained and developed? (2) How does personal and social media affect the way face-to-face social events are organized and coordinated? and (3) How does personal and social media influence the organization of travel activities?

In the next section, we lay out our theoretical approach before briefly setting out the data we draw on in this study. We then move on to present our findings before closing with a discussion on the potential short and longer-term implication of personal and social media use for travel. Overall, we argue that personal and social media have become central elements in urban dwellers' leisure travel practices, both before, during and after social events, and in in their travel there and back. We relate this to a wider change in the way social networks are developed and sustained, in particular among younger individuals. We argue that the diffusion of personal and social media in everyday travel influences the coordination and content of leisure trips and leisure activity itself.

Theoretical framework

Networked individualism

Personal media is involved in an ongoing transformation, both in the way individuals and communities communicate and develop interpersonal relationships and in how they form and maintain social networks. For Castells, this stems from a structural shift, where increasing use of web-based communication in everyday life is leading to the emergence of a "networked individualism" (Castells, 2003, pp. 130–131) where

> we have entered not only a new technological paradigm, but a new form of organizational structure for everything we do . . . from the vertically organized, standardized, rationally structured, hierarchically structured forms of activity to networking forms of activity.
>
> (Castells, 2000, p. 152)

His argument follows that Internet-based technologies have wrought a new social ontology where social relations and interactions are now organized collaboratively. In this, he holds organizational activities to be distributed equally across the members of a social network. Following Castells (2000, 2003), Rainie and Wellman (2012) argue that as communication becomes increasingly mediated and individualized, a shift in sociality is taking place, too, where we

> have become increasingly networked as individuals . . . it is the person who is the focus: not the family, not the work unit, not the neighbourhood, and not the social group.
>
> (Rainie & Wellman, 2012, p. 6)

In short, they argue that interactions between individuals are now primarily mediated and carried out across horizontally organized social networks. In contrast to earlier more place-based networks, both Castells and Rainie and Wellman assert that individuals are now more active in developing and sustaining their own personal social networks. Thus, personal media spurs a development toward networked individualism where individuals grow their own personalized social networks with a wide array of social relations. For Wellman (Wellman, 2002; Wellman et al., 2003), the shape of such networks rests on a differentiation between local and global and in the strength of ties between social network members.

In contrast, some theorists hold Castells's position to be techno-determinist and structuralist (Garnham, 2004). Meanwhile, others agree that transformations to social organization are taking place through the use of Internet-based technologies and personal media but reject the a priori assumption that technology is central to it. Instead, they tend to hold a less radical position on the degree to which social ontology has changed and draw comparisons with a similar shift over a longer durée through earlier media (Webster, 1995). To that end, while we do not expand on social network theory literature in this chapter, we do draw on "social network" and "networked individualism" as useful concepts in our analysis and discussion of findings.

Structuration theory

Whilst we draw on social network theory, it proves limited in its capacity for addressing how travel and communication practices relate to social structure developed by communication activities. To that end, we also draw on sociological theory surrounding the emergence and construction of social structure, in particular structuration theory (Giddens, 1984). We do not hold mediated communication, physical mobility (travel) and face-to-face meetings to be separate processes. We link them through the increasing the personalization of media and an increasing level of mobility in everyday life. We also hold that it is communication that activates and sustains social networks of relations, the core of which are face-to-face meetings – which usually revolve around certain tasks, activities or ideas. In this, travel holds an important function in how individuals develop and sustain social relations. Doing so is an ongoing activity, and social networks only function if recursively activated through the intermittent co-presence of some of its members (Feld, Suitor, & Hoegh, 2007; Urry, 2007). If all members leave, the social network fails to exist. We also understand travel and communication as dynamic elements in the ongoing reproduction of social structure. In essence, social networks are an array of practised relationships and links between individuals that collectively constitute social structures.

In accordance with structuration theory, social organization emerges through a structure/agency duality where neither structure nor agency is a priori. Instead, as individuals go about their daily lives, they draw on past "memory traces" (Giddens, 1984, p. 45) to negotiate various "rules" and "resources". That is, individuals' practices are structured by the shared rules and resources they encounter and which they understand through reference to past biographical experience. At the same time, their practices serve to reproduce or challenge that structure. In this, practices are held as "*both the medium and the outcome of structure. . .*" (Giddens, 1984, p. 27).

Thus, social networks may be seen as "schemata", or collective memory traces, that structure social systems over time and space (Corman & Scott, 1994). For example, travelling to meetings and day-to-day communication are both mundane activities that reproduce and maintain the structure of a social network whilst also being structured by it. In almost any social group, there are expectations and norms (rules) for attending meetings and events. Likewise, as Urry (2007, p. 221) notes, the sustainment of social networks is rarely free; it relies on individuals' use of "network capital". In this, it is individuals' use of trains, cars, buses, trams, aeroplanes and personal media (as resources) that make it possible for them to sustain and develop wide-ranging personal networks. To extend this, we draw on Corman and Scott (1994) who propose that such network structures are activated when individuals take part in common or shared activities, or "foci", which are then enacted through various "triggering events".

In this, travel to and from face-to-face meetings and communication through personal media can be considered parallel (although tightly integrated) activities which are both necessary to keep the network alive (see Figure 7.2).

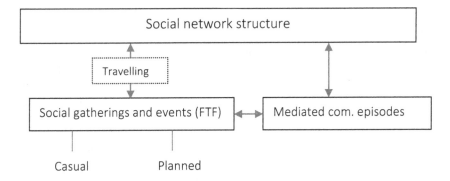

Figure 7.2 Conceptual overview and relations

Note: FTF = Face-to-face.

Another key concept we adopt from structuration theory is "routinization". Routines and routinized social practices that relate to communication and face-to-face meetings are crucial in building and sustaining social structures and in framing the actions of networked individuals. As individuals go about their daily lives, their routine practices include interaction with others with varying degrees of "closeness", for example family members, work colleagues, unknown commuters on the same daily train journey and social media acquaintances. These routinized interactions constitute and sustain social networks. At the same time, social networks serve to constitute the local cultural schemes that frame individual travel practices. In this, routinization may be described as the process whereby structures (specifically an ordered set of rules and resources) become meaningful in duality. As individuals go about their daily lives, they recursively constitute structures (and social networks). They also constitute predictable patterns of activity (routines), which, in turn, provide a sense of continuity – both in the predictability of the world and of self-identity.

Recently, a stream of general practice theories have begun to elaborate on the role of social practices at an ontological level (Pantzar & Shove, 2010; Reckwitz, 2002; Warde, 2005), often drawing on structuration theory to do so (Postill, 2010). In general, practice theories reconcile a macro-sociological understanding that structures coerce and constrain individuals' actions, for example the economy, a legal system, religion and so on, with a micro-sociological understanding that structures are constructed through individual interactions (Ortner, 2006; Postill, 2010). They do this by holding individual agency and structure as recursively co-constituted into routine practices (Giddens, 1984). Epistemologically, macro-sociologists tend to treat individuals as passive actors coerced or rationally motivated to act in some way. Meanwhile, micro-sociologists tend to assume that individuals are either passive cultural dupes or knowledgeable of the norms they reproduce or challenge (Ortner, 1984). By contrast, practice theorists assume that individuals are reflexive and constantly self-monitor their own action as they go about everyday life but do so for the most part without direct rationalization.

Social events and travels

As discussed earlier, social events, meetings, visits, sports, holidays and various other leisure activities are crucial for developing and sustaining personal social networks. They also feed directly into the constitution of social structure. Although social events may be mediated, conducted through digital media in virtual social spaces, there is little evidence to suggest they could replace face-to-face gatherings entirely (Aguiléra, Guillot, & Rallet, 2012; Fiore, Mokhtarian, Salomon, & Singer, 2014). This is not to say the only reason for social events is to sustain and develop networks, but it is an important implication and consequence. More directly, individuals travel to face-to-face events and meetings to receive social support, accumulate social capital, and to increase their subjective well-being. (Larsen et al., 2008; Vos, Derudder, Acker, & Witlox, 2012). This is much in line with a key assumption in traditional micro-sociology, emphasizing the importance of social gatherings and co-presence for development of common meanings, social norms and identity (Garfinkel, 1967; Goffman, 1967). Following Randal Collins (2004), co-present social gatherings represent interaction rituals that generate emotional energy, solidarity, and feelings of morality. People's lives are structured by such "chains of rituals", and they are crucial for our development as social human beings. Thus, it is reasonable to see social events as a crucial element in an understanding of social interaction and travel activities.

To aid our analysis, we plot social network activities on a temporal axis, using five different "temporal stages" (see Figure 7.3). This enables us to compare personal media use and travel practices: before an event takes place, during the event, on the journey to and from the event and after the event itself. As we argue in the following, social media is used actively along all five phases to sustain and support social networks.

Methods

We base our argument in an analysis of 16 in-depth interviews with residents living in Oslo, Norway. Seven interviews were carried out in October 2015 as part of a COST Action (TU1305) short-term scientific mission. A further nine interviews were conducted in August 2016 in collaboration with the University of Oslo. We recruited interviewees through a targeted process involving social media (Facebook and Twitter) and through personal contacts. This ensured a representativeness of informants at different life stages and with different experiences of using media technologies.

About half of the sample are under 30 years of age. A similar range are aged between 30 and 60, with only two interviewees aged 60 or older. However, we make no claims that our sample represents any wider population; it is purposeful and non-representational (see Table 7.1).

Each interview lasted for about 1 hour, and a semi-structured questionnaire was used. The interviews centred around the use of personal and social media

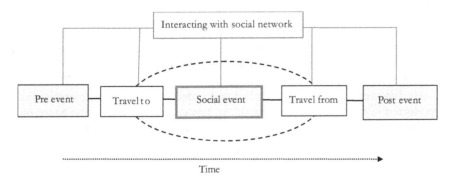

Figure 7.3 Social networks and use of personal media in relation to travel to and from social events

Table 7.1 Anonymized interviewee details (sample)

Pseudonym	Sex	Age	Occupation	Period
Carl	Male	51	Self-employed	Nov
Adam	Male	48	Full-time employed	2015
Barbara	Female	63	Retired	
Nicole	Female	27	Full-time employed	
Sebastian	Male	32	Full-time employed	
Sarah	Female	29	Full-time employed	
Jennifer	Female	36	Self-employed/Maternity	
Bjørn	Male	19	Student	Sep
Alf Jørgen	Male	21	Student	2016
Oda	Female	22	Student	
Elise	Female	24	Student	
Dan	Male	20	Student	
Annbjørg	Female	72	Retired	
Frida	Female	22	Student	
Martin	Male	36	Full-time employed	
Kai	Male	40	Full-time employed	

during travel, and each interview participants were presented with an opportunity to complete a personal network diagram (see Figure 7.4). This socio-spatial network diagram was designed to capture important friends and family member and their location/distance to the informant (ego). Locational categories included within household, in the same city/region, within the country or abroad. The informants were encouraged to talk about the historical development of the relationships and how often they usually communicated face-to-face or through media. The informants were then first asked to explain their day-to-day contact and interactions with people specified in the network diagram. This involved mostly informal meetings and gathering over the day, how this was arranged for, the transport mode used and the role of mobile media. Second, we asked for travel and communication related to more regular and routinized gatherings, such as weekly meetings at the gym, regular trips and so on. Third, we asked for communication and mobility related to less routinized social events, such as going to concerts or joint holidays abroad. Finally, we asked more generally about the development of media use habits and what impact this had had on their social network and their travel habits.

All data were recorded, transcribed and coded in NVivo. The analysis was conducted in two rounds (in line with the data gathering), and it was based on codes structured by the overall theoretical approach and questionnaire (Patton, 1987).

Findings

In this section we present key findings from our interviews. In accordance with the temporal stages identified earlier, we focus on how personal media are used before, during, and after events and on the way to and from each event (see Figure 7.3). At each temporal stage we draw together key themes, relating them to our research questions in order to address any emerging changes to social networks or how this affects the coordination and organization of travel activities.

Pre-travel social media use

Coordinating shared trips: There is often a stage of intense social interaction and coordination amongst members of an existing social network prior to an event; for example friends may phone each other to set a time or place to meet or discuss travel plans for getting there. However, personal media (including social media apps) appear to allow horizontally organized ad hoc social networks to form. For example, when Sebastian arranges a car-sharing/car-pool trip from Oslo to Gothenburg, he describes it as an activity coordinated amongst strangers using social media to establish a social network. By using a Facebook group, Sebastian makes contact with others (often other migrant Swedish workers) to arrange the specifics of the journey. In this, individuals that are unknown to each other will share a car and travel together from one country to another:

> I use Facebook for trying to use car pools . . . in Oslo, there is a very big Swedish population that travels a lot back to the home country . . . one group on Facebook simply called "car pools going to and from Oslo" and people just put on there: "I am going to Gothenburg" . . . usually the driver says "I am going to there" and then it's "how much do you want to take me there?", and it's usually 20 [NOK] or something.
>
> (Sebastian)

On the surface, Sebastian's account suggests a shift toward networked individualism, where he (and others) form a travel-based social network horizontally through a social media app. However, he later adds that "[p]*eople with cars are more in demand than without. . .*" (Sebastian). This suggests a deeper nuance and inequality in the social network which disrupts any directly horizontal structure. Where some individuals (drivers) hold more control over a key resource (the car) than others, the social network appears to be hierarchical. He also describes a gendered division in the acceptance or rejection of individuals from a car-pooling social network, where "[f]*emales get more responses than males. . .*" (Sebastian). This suggests structural differentiation also plays a part in structuring social networks. In short, what Sebastian describes is a form of networked individualism in the organization of a shared travel activity through a social media platform. However, he describes it as unequally structured, in terms of both access to resources and structural differentiation amongst social network members. In doing so, he opens a question on how travel-based activities organized by a social network across social media use might differ from those of traditional sets of social relationships such as family units or groups of friends.

Organize journeys and social events: Exploring how family members coordinate to arrange meeting at an event highlights a difference between personal and social media use amongst different types of relationship and the social organization each can foster. For example, when Martin explains how he uses social media to arrange meeting family and friends at Christmas, he describes a horizontally organized social network. Organizational tasks are distributed evenly amongst family members and friends, with Facebook groups used as a communication channel to mediate the communication of a pre-existing social network:

> The Christmas table with the boys is a yearly happening . . . these parties are organized through Facebook, a group, and coordinated there. Everything is set a long time in advance. Everyone is part of the group and comes up with suggestions and so on.
>
> (Martin)

In contrast, when Nicole describes how her family arranges to meet at Christmas, she highlights a deep inequality. Her family members do not engage with personal or social media, leading her to take a lead role in organizing the meeting and all associated travel:

> [M]y father doesn't know how to use a computer, even though he is not that old, and my Mum is not that good at using the internet. My sister neither – and that's kind of strange, but she isn't. So, I just kind of do all the planning . . . I take all the recommendations off them, and the dates and then I just book the hotels or whatever.
>
> (Nicole)

Nicole's account covers an uneven use of personal media, social media and the Internet amongst family members. In turn, her account follows Sebastian's in presenting a nuanced inequality in the social organization of a travel-based activity. However, their different accounts open a question on the timescale involved with organizing such activities. For Martin and Nicole, arranging to meet family at Christmas is something they plan and coordinate in advance. For Sebastian, carpooling may be planned in advance or organized with a relative degree of immediacy. In turn, this requires a better understanding of the social organization involved with more immediate events, to explore how it might differ from planning ahead, and the way in which social networks are practised in both.

Immediate planning for informal gatherings: When individuals make ad hoc arrangements to meet for an informal gathering at short notice, they tend to use a range of personal and social media. For example, when Alf-Jørgen arranges to meet (face-to-face) with his friends, he explains that his personal social network use SMS text messages Facebook's Messenger app to plan immediately prior to meeting:

> We always plan before meeting. There are no systems, so there is always talks on SMS or Messenger before we go out, to agree on where and when. Not so much during or after.
>
> (Alf-Jørgen)

However, Alf-Jørgen notes that after setting the plan, there is little communication among social network members until they meet in person. Interestingly, Oda adds that immediate plans to meet with her social network often overlap with advanced planning, with frequent micro-coordination over SMS text messaging and various social media services:

> There is absolutely a lot of planning in advance. It may happen that we randomly meet at the library and decide to sit and read together, but usually we send messages a couple of days before or the day before. Mainly the planning is done over SMS but also Snapchat and Facebook.
>
> (Oda)

In short, Oda's account suggests that organizing an informal and relatively immediate activity (meeting friends at the library) can involve a process of negotiation between members of a horizontally organized social network. In this, Oda presents personal and social media as important resources in structuring the practices of her social network. Comparing this with the previous accounts, Oda's example suggests that personal and social media clearly have an impact on how events are planned and organized.

Our impression is that their use means social events are more collaboratively planned than ever before and that random visits have become articles of virtue. In this, groups and communities that use social media can function as "information hubs", where information sourced from elsewhere on the Internet may be distributed and shared amongst the social network, for example timetables, hotels, opening hours and so on. We also find that pre-event communication is often focused on negotiating and setting the details of an activity (the how, where and when of an event), which can, in turn, structure the planning of subsequent trips. However, pre-event communication can also be seen as a reflection of group identity or as delimiting boundaries with other groups. However, our argument at this stage is limited in addressing these points. We have only covered the first of five temporal stages. This leaves open questions about how social and personal media are used during travel to and from an event and how this relates to social organization and the coordination of shared activities.

Social media use to and from events

Surveying (social) information: In exploring personal media use during travel we find that when individuals use personal or social media to plan or organize meetings/events, they tend to use the same personal or social media during their travel to it – especially on public transport. However, social media use while travelling often relates to "modi", ranging from active to passive observations, and can also relate to reflection on previous events. For example, in terms of personal media use during travel, our interviewees describe browsing through social media content uploaded by friends or acquaintances to keep up-to-date with social information and activities. We refer to this as "surveying social information". They also use personal media when travelling to access general news, business information and travel-related information and for communicating with others. Rather than separating these as distinct activities, our participants tend to seamlessly integrate them into a broader practice of personal media use. For example when Sarah describes her use of a smartphone during a routine daily commute to work, she attends to both her work and personal emails at the same time, whilst also reading the news and checking Facebook:

> I always have the phone with me . . . on the bus on my way to work in the morning I would look through my job e-mail, my personal e-mail, newspapers, and Facebook to get updated, and then during the day I would use my laptop both at work and probably also after when I get home.
>
> (Sarah)

Alongside this, Sarah also depicts continued social media use via a laptop during the day to survey (social) information whilst at work. In turn, this suggests a need or want for an ongoing connected presence (Boase, 2008). However, much of the social information she consumes has been "produced" by members her existing personal social network. In turn, her account opens a question about the extent to which information communicated across social media might structure the actions of individuals without any direct interaction; she surveys information but does not necessarily communicate directly with those producing it. At this point, it is also worth noting that the specific choice of social media does not significantly change the way it is used. For example while Sarah attributes her use of Facebook to surveying social information, Bjørn depicts a similar use of Twitter:

> I use my mobile phone all the time. Mainly to entertain myself or check information that I am curious about. Twitter is to see what's going on, get first-hand information from people that are central in areas of interest for me.
>
> (Bjørn)

In this, Sarah and Bjørn both depict an individualized form of social media use whilst travelling in order to keep up-to-date with their personal social network. In turn, their accounts open a question on how keeping up-to-date via social media when travelling might relate to the shape of a social network, or to the distribution of activities in collaborative organization of an event.

Grooming of relations: Alongside browsing through social media to survey social information, our participants also describe using personal media when travelling to communicate with other members of their personal social network. For Oda, this is a routinized practice that directly correlates with her maintenance of a connected presence:

> [F]or me it has become so constant and omnipresent . . . part of the everyday entertainment to have the phone in the hand. Just to scroll down the Facebook feed to see what's there turns into communication with my

network, by clicking likes and comments, although I cannot count that as real communication.

(Oda)

In short, Oda presents her personal media use as a form of entertainment that somehow differs from "real communication". Her assertion resonates with a sentiment amongst a wider group of highly active social media users who attribute personal media use when travelling as a means of countering boredom. For some participants, it also represents a continuous grooming of social relationships. For example, when Yvonne uses her smartphone on a long train journey, she describes her active use of all the social media service apps on her smartphone. Like Oda, she holds this to be a form of entertainment. She also attributes it to surveying social information and grooming social relationships:

During the 12-hour trip, I used every application on my mobile. I had to go through Snapchat and everything, I hate travelling by train. I constantly sat and sent Snaps, talked on Messenger and Facebook, but didn't use the phone. Sent messages, went through Instagram, VSCO and all there is of pages on the net.

(Yvonne)

In part, the ability to use personal media and social media apps continuously throughout a long journey marks a key difference between the smartphone and prior modes of communication. An important dimension of communicating over social media is that the communication piles up to long conversational streams with no apparent beginning or end. For Alf-Jørgen, this is an attractive feature that simultaneously entices him to use social media whilst making them omnipresent and harder to stop:

I think communications gets more high-frequent. You cannot send text message continuously, but a chat on Messenger can go on for hours. . . . You just send a link, it is less binding, free for any obligations.

(Alf-Jørgen)

In short, not only are social media used during travel to alleviate boredom as a means of surveying social information and grooming social relations; they also provide a constant stream of communication amongst social network members. In turn, and following on from the discussion of personal and social media use, prior to travel (to plan or organize) this opens a question on how they might also be used in micro-coordination of activities amongst a social network during travel to or from an event.

Micro-coordination: The portability of contemporary personal media provides a set of affordances for wayfinding and for coordinating where to meet whilst en route to an event. In most cases this is in addition to earlier organization of the event. For example when Bjørn describes going to a theatre, he refers to having previously arranged to meet friends at a specific bookstore. He also describes the micro-coordination involved amongst social network members whilst travelling there:

> [W]e used social media to share arrangement info and communicate where we were, where the others were, how far away they were from Tanum [a bookstore] – where we planned to go. . .
>
> (Bjørn)

In this, Bjorn suggests that personal media changes travel practices insofar as individuals are better informed on the exact location of social network members during travel to an event, a point the underlines the connected presence. Similarly, when Daniel describes how he found his friends on a public transport carrier, he reveals a similar type of micro-coordination. He and his friend arranged to meet in advance in order to go run in a park outside Oslo. Whilst on the way, they used social media to coordinate sitting on the same train and wagon/carriage:

> He comes from Bøler, about 40 minutes away. We try really hard to get on the same train, he can text "I am now at Majorstua", or "I am on the National theatre", so I have to estimate the right stop for me to get on to the same train. We try to run every Tuesday, but if it changes we move it and update each other through messenger.
>
> (Daniel)

In this, Daniel depicts his social media use as instrumental, both in arranging the activity and in synchronizing the mobility of social network members whilst on the way there. This opens a question about whether the micro-coordination of activity during travel (via personal media) alters sociality or how social networks members interact with one another.

For Sebastian, using social media to establish a new ad hoc social network specifically for travel-based practice can blur boundaries between friends and acquaintances. For example, some carpooling trips lead to an ongoing social relationship between social network members where social media acquaintances that meet for one carpooling journey may contact each other (as friends) to arrange another journey. That is individuals tend to re-enact the same personal social network when triggered by the same event:

> I have one, well I suppose I can kind of call her a friend . . . I got a lift off her to Gothenburg from Oslo, and then we talked, and then "oh, you're going to Gothenburg again, can I come?" . . . We have developed, like "meet me in Oslo at 2 o'clock".
>
> (Sebastian)

In turn, this challenges any notion of the social networks enacted through social media being ephemeral. Instead, we argue that social network ties can sit in latency until re-enacted (or retriggered). Interestingly, Sebastian also notes that carpooling (as a travel-based practice) has an established set of rules and a further hierarchical structure (a sociality), with specific tasks assigned to individuals based on their position in the car:

> [T]here is like an unwritten, well the rule is kind of written now. Your job as the driver is to get us there safely. Then your job is to be awake and talk to the driver. It's very rude to fall asleep when you're the shotgun next to the driver. You have to be awake, talk to the driver, and possibly DJ duties. The people on the backseat, well, they can sleep, but that could be considered rude as well. Their job is to dispense the snacks and remind the driver there are such things as bathrooms . . . also if there is a lot of luggage, then it is on them.
>
> (Sebastian)

In combination, what Alf-Jørgen, Bjorn, Oda, Sebastian, and Yvonne all demonstrate is that personal media has led to the formation of new travel-based practices being carried out whilst in transit. This includes surveying social information and ongoing communication to refine plans as part of a connected presence and the increased micro-coordination of activity. However, a focus on how personal media are used prior to, and during travel to and from events speaks little of how they might impact on the way activities are carried out at the event.

Social media use while at an event or meeting

Sharing the moment with distant network members: Individuals often use social media for continual micro-coordination at events, both to communicate with co-present social network members and to include others who are not physically present. For example, when Oda attends a party with her mother and her friend (Hanna), she uses a range of social media apps alongside SMS text messages to arrange plans for the event and to organize how to get there and then continues her use whilst at the event. Once there, she uses Snapchat to send pictures to her boyfriend, who is not physically present:

> I sent a bunch of Snaps [Snapchat messages] to my boyfriend during the party since he was not there. I communicated with Mummy through Facetime and SMS, and Hanna and I used Messenger and Snapchat to talk about the plans.
>
> (Oda)

What Oda demonstrates is that personal media use extends beyond the planning of an event or organizing how to get there and on to the ongoing maintenance of a personal social network whilst in situ. In Oda's case, including her boyfriend in the event through Snapchat messages and images provides him with a virtual and continual presence at the event. In turn, this opens a question about how social networks might be sustained during longer journeys or events, where a continually connected virtual presence is not feasible.

Sustaining network while on longer journeys: For some participants, using social media to sustain and develop social networks during longer trips, either to other parts of the country or abroad, provides a very different context for their use than in examples of more immediate events. For example, when Elise discusses her holiday, she describes often being alone and self-critically reflects that a substantial part of her time away involves making contact with friends at home:

> I don't go anywhere without my mobile, even on my cottage when I have holiday . . . In a way it has captured a lot of my life. Mostly I sit on the mobile just browsing through Facebook, then moving over to Instagram, moving over to Snapchat and then Facebook again. I will say it is getting close to a sickness . . . There is a need to be in contact with people all the time, to share things. When I was in Lyon last year it was there all the time: take up the phone, take pictures, send out, put on Snapchat. You have to share all the time instead of just sitting and enjoying on your own, you just need to be in contact all the time.
>
> (Elise)

What Elise highlights is that physical absence from a local social network and the lack of opportunity to participate in face-to-face events can trigger a strong need to survey social information and a need to a maintain a continually connected presence. Similarly, the need to remain connected with a social network throughout a long-term physical absence is something Frida discusses at length in her blog, taking part in about a student exchange programme:

> After secondary school I went to Australia for a year, and I used Facebook, Instagram, and also a blog. Through the blog, I communicated with my readers what I was doing. The[n] Snapchat came – so now I use Facebook for planning, Snapchat for daily talk and Instagram for looking at picture[s] and entertain[ing] myself.
>
> (Frida)

In this, Frida notes that not only is social media used to keep in contact with others when absent from local face-to-face events but that such absence also triggers a higher-frequency of social media use. This, at times, is juxtaposed with an expanding personal social network – where long-distance stays abroad also generate new connections and altered network structures. When exploring informants' personal networks, it became evident that younger informants tended to have a rich network of friends, with subgroups of friends located in other parts of the country or abroad. In this, we identify two types of change in their lives that provide long-distance ties: relocation from home to another town (usually related to education) and longer-term visits abroad for extended holidays and/or travel. For example, when Oda discussed her student exchange to study at a university, she highlights how it transformed her social network:

> My exchange in France and Spain has changed my social network in many ways. It became clear to me whom I really had strong ties to in Oslo and

many weaker and trivial friendships fell away. At the same time, I got to know a number of new people down there, obviously.

(Oda)

Expanding on this, she provided a personal network drawing (an ego-network diagram) to illustrate her understanding of the new and more dispersed personal social networks that her experiences have generated (see Figure 7.4).

What her account demonstrates is that time abroad expanded her personal network, providing her with a membership to various different local social networks and a plethora of latent connections. In turn, her connection between an increasing frequency of social media use to maintain a connected presence in an expanded set of social networks suggests a greater frequency and volume of use over time and that a longer trip away requires extensive use during the event. However, this leaves the final temporal of stage unaddressed, requiring an exploration of how personal and social media are used after an event and of how that relates to travel-based activities.

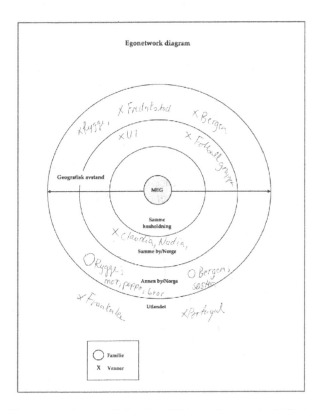

Figure 7.4 Oda's personal network drawing. Distance from centre indicates geographical distance from ego (household, same city, other domestic city, abroad). O = family; X = friends.

Post-event use of personal and social media

Collective sense-making and maintenance of the social networks: Overall, we find that after an event, participants' attention soon moves on to planning the next event. However, there are some instances where personal and social media are used to share picture and to reflect on an event (both individually and collectively) in order to continually maintain the social network by reference to a shared history or shared set of memory traces. While Sebastian touched on this earlier in his discussion of repeated interaction with the same social group to arrange carpooling, our participants' accounts of personal and social media use after an event typically involved maintaining a social network by sharing photographs, messages and comments about family dinners, celebrations and concerts. For example after returning from an event related to her father's 50th birthday party, Oda explains how her social network used Facebook to share pictures of the event and to keep in continual contact:

> Afterwards there were a bit of SMS back and forth, congratulation each other with a successful organized event. And pictures on Facebook, a lot of pictures on Facebook.
>
> (Oda)

What this indicates is that alongside the higher frequency of social and personal media use required during an event to continually groom social relations and to maintain presence when physically absent from local face-to-face events, such use also continues immediately after an event. Whilst this evidences an extended form of social network maintenance, at a theoretical level, it also indicates that social media-based interaction captures some of the collaborative "sense-making" of a social group, for example in the collaborative construction of a shared interpretation of the meaning and significance of an event, which in turn informs the future practices of social network as a set of shared memory traces.

Discussion and conclusion

In this chapter, we have demonstrated that personal and social media play a key role in how individuals coordinate and organize social events and in their ongoing maintenance of social networks. In this, personal and social media has become a central element in various social practices, both before, during and after social events, and in transit to or from them. As the interviews suggest, how personal and social media are transforming travel-based social practices varies. Initially, we raised a question on how personal and social media technologies influence the development and sustainment social relationships. Through our research we find three trends, all of which have already discussed in earlier literature to some extent: first, we noted that social interaction has become more communication-intensive through the affordances embedded in personal and social media. This increased intensity has not emerged overnight. Earlier theorists have already argued that mobile media and the Internet have spurred a growth of "connected presence" (Katz & Rice, 2002; Licoppe & Smoreda, 2004). However, more recent advances in social media web applications seem to have enhanced an ever stronger "always on" mode of inter-action amongst dispersed social networks to "keep in touch" – a feature that cuts across a wider range of activities and social situations. For younger participants especially, the threshold for communication appears to be very low with a con-tinuous flow of communication as a typical feature of social network sustainment. This higher frequency of interaction also continues and is sometimes even intensi-fied when people are temporary geographically distant from their immediate social network. In this, communication practices appear to be moving toward a real-time mode, where mundane pictures of objects encountered in the flux of everyday life, spontaneous reflections, discussion and web stories of experiences are continually produced and circulated as part of the maintenance of social networks.

Moreover, the social norms for communication within a network also seem to be drifting towards expectations of more actively mediated communication to keep the social network alive. Similarly, there are explicit expectations that close friends, partners and family members will maintain their relationships through social media platforms. In short, not only does interaction with others sustain and develop social networks; it also seems to demand more frequent interaction over social media alongside face-to-face interaction, with both equally critical parts of the sociality of a social network. Being outside the communication risks social exclusion, especially for younger participants.

Another key trend is that the social networks, in general, have a high level of diversity and usually contain local, as well as distant, connections. Trips and stays abroad are important factors that not only contribute to this, but also ties forged in childhood and youth have, in many cases, been continued and sustained. Hence, the networks seem to be stretched out in both a geographical and temporal sense. At this point, our findings seem to be aligned with networked individualism. It is also striking how social relationships were recognized as important by partici-pants, and as something to be worked on by using social media. Their awareness of their social network as a resource with social value was relatively strong. Through our exploration of the entanglement of personal and social media, and ICT more broadly, in travel we found a shift towards more horizontally organized networks (at the level of ontology) through personal and social media. However, turning specifically towards travel activities, this shift seems to have influenced partici-pants' leisure-based travel activities along a different set of dimensions. There were no indications that mediated interaction is beginning to substitute face-to-face meetings and gatherings. Instead, mobile communication and personal and social media use represent an additional layer of communication employed at all temporal stages of a social event. Similarly, social networks with a high number of distant ties (physically absent members) may lead to new sets of obligations and needs for social support and well-being. As such, the active use of social media is likely to initiate more travel to face-for-face meetings, not less.

There was also a clear trend towards active use of personal and social media to plan trips and informal meetings, even local ones. For example, routinely meeting at distinct places (cafés, street corners, etc.) was often replaced by a coordination of events through media. At the same time, in the context of mundane everyday events, "micro-coordination" increased amongst small groups (two or three people) as the event approached, for example to arrange meeting on the same train, or to define the exact meeting place. For less mundane events involving large social networks (families, groups of friends), participants often described their use of personal and social media to coordinate the activities as a routinized practice. For example Martin and Nicole earlier both described their arrangement of a family meeting at Christmas as a typical event where geographically distant family members coordinate to discuss travel activities as well as food, presents and various other issues. Portrayals of these larger social networks represented personal and social media as important hubs for sharing information. In this, we identified a trend toward horizontally organized networks where personal and social media were engaged as platforms to organize shared transport resources, for example the sharing of cars in carpooling. We argue that these new network structures are key resources in the organization and coordination of both local and distant travel activities. Another central dimension of our research relates to the experience of being on the move. We found that the time used for travelling was important for participants, in both their active and passive communication and for interacting with their social network. In short, travelling has become a setting for grooming and observing social ties through personal and social media. Arguably, this may lead to a hypothesis that the release of travel time for social activities makes it possible to sustain a wider and stronger network of relationships. In any case, it seems clear that the new media landscape of ubiquitous personal and social media makes it possible to sustain social networks that formerly centred on co-located activities in the household or at workplaces or at the schoolyard whilst being on the move. This makes the organization of social networks more efficient for individuals, even as the boundaries between various social situations and networks become less clearly defined.

References

Aguiléra, A., Guillot, C., & Rallet, A. (2012). Mobile ICT and physical mobility: Review and research agenda. *Transportation Research Part A, 46*, 664–672.

Alexander, B., Ettema, D., & Dijst, M. (2010). Fragmentation of work activity as a multidimensional construct and its association with ICT, employment and sociodemographic characteristics. *Journal of Transport Geography, 18*, 55–64.

Axtell, C., Hislop, D., & Whittaker, S. (2008). Mobile technologies in mobile spaces: Findings from the context of train travel. *International Journal of Human-Computer Studies, 66*, 902–915.

Baym, N. (2010). *Personal connections in the digital age*. Cambridge: Polity Press.

Billinghurst, M., Reichherzer, C., & Nassini, A. (2016). Collaboration with wearable computers. In B. Woodrow (Ed.), *Fundamentals of wearable computers and augmented reality* (2nd ed., pp. 661–680). Boca Raton, FL: CRC Press.

Boase, J. (2008). Personal networks and the personal communication system: Using multiple media to connect. *Information, Communication and Society, 11*(4), 490–508.

Cairncross, F. (1997). *The death of distance: How the communication revolution will change our lives.* Boston, MA: Harvard Business School Press.

Castells, M. (2000). The contours of the network society. *Foresight, 2*(2), 147–157.

Castells, M. (2003). *The internet galaxy: Reflections on the internet, business, and society.* Oxford: Oxford University Press.

Castells, M., Fernandéz-Ardévol, M., Qiu, J., & Sey, A. (2007). *Mobile communication and society: A global perspective.* Cambridge, MA: Massachusetts Institute of Technology Press.

Chen, W., & Wellman, B. (2010). Net and jet. *Information, Communication & Society, 12*(4), 525–547.

Choo, S., & Mokhtarian, P. (2005). Do telecommunications affect passenger travel or vice versa? Structural equation model of aggregate U.S. time series data using composite indexes. *Transportation Research Record, 1926,* 224–232.

Christensen, T. (2009). 'Connected presence' in distributed family life. *New Media and Society, 11*(3), 433–451.

Collins, R. (2004). *Interactional ritual chains.* Princeton, NJ: Princeton University Press.

Corman, S., & Scott, C. (1994). Perceived networks, activity foci and observable communication in social collectives. *Communication Theory, 4*(3), 171–190.

DaCosta, F. (2013). *Rethinking the internet of things: A scalable approach to connecting everything.* Berkeley, CA: Apress (Open).

Feld, S., Suitor, J., & Hoegh, J. (2007). Describing changes in personal networks over time. *Field Methods, 19*(2), 218–236.

Fernandez, Z., & Usero, B. (2009). Competitive behavior in the European mobile telecommunications industry: Pioneers vs. followers. *Telecommunications Policy, 33,* 339–347.

Fiore, F., Mokhtarian, P., Salomon, I., & Singer, M. (2014). 'Nomads at last'? A set of perspectives on how mobile technology may affect travel. *Journal of Transport Geography, 41,* 97–106.

Garfinkel, H. (1967). *Studies in ethnomethodology.* Englewood Cliffs, NJ: Prentice Halls.

Garnham, N. (2004). Information society theory as ideology. In *The information society reader* (pp. 165–184). London: Routledge.

Giddens, A. (1984). *The constitution of society.* Berkley; Los Angeles, CA: University of California Press.

Goffman, E. (1967). *Interactional ritual: Essays on face-to face behaviour.* Garden City, NY: Doubleday, Anchor Books.

Gössling, S. (2018). ICT and transport behavior: A conceptual review. *International Journal of Sustainable Development, 12*(3), 153–164.

Gripsrud, M., & Hjorthol, R. (2006). Working on the train: From 'dead time' to productive and vital time. *Transportation, 39*(5), 941–956.

Igarashi, T., Takai, J., & Yoshida, T. (2005). Gender differences in social network development via mobile phone text messages: A longitudinal study. *Journal of Social and Personal Relationships, 22*(5), 691–713.

Jensen, O., Sheller, M., & Wind, S. (2015). Together and apart: Affective ambiences and negotiation in families' everyday life and mobility. *Mobilities, 10*(3), 363–382.

Julsrud, T., & Bakke, J. (2008). Trust, friendship and expertise: The use of email, mobile dialogues and SMS to develop and sustain social relations in a distributed work group. In R. Ling & C. Campbell (Eds.), *The mobile communications research annual: The reconstruction of space and time through mobile communication practices* (Vol. 1). New Brunswick, NJ: Transaction.

Julsrud, T., & Denstadli, J. (2017). Smartphones, travel time-use and attitudes to public transport services: Insights from a study of urban dwellers in two Norwegian cities. *International Journal of Sustainable Transportation, 11*(8), 602–610.

Jurgenson, N. (2012). When atoms meet bits: Social media, the mobile web and augmented revolution. *Future Internet, 4*(4), 83–91.

Katz, J., & Rice, R. (2002). *Social consequences of internet use.* Boston, MA: Massachusetts Institute of Technology Press.

Kenyon, S., & Lyons, G. (2007). Introducing multitasking to the study of travel and ICT: Examining its extent and assessing its potential importance. *Transportation Research Part A, 41*(2), 161–175.

Kim, H., Kim, G., Park, H., & Rice, R. (2007). Configurations of relationships in different media: FtF, email, instant messenger, mobile phone, and SMS. *Journal of Computer-Mediated Communication, 12*(14), 1183–1207.

Kim, J., Rasouli, S., & Timmermans, H. (2018). Social networks, social influence and activity-travel behaviour: A review of models and empirical evidence. *Transport Reviews, 38*(4), 499–523. doi:10.1080/01441647.2017.1351500

Kraut, R. (1998). Internet paradox: A social technology that reduces social involvement and psychological well-being? *American Psychologist, 53*(9), 1017–1031.

Larsen, J., Urry, J., & Axhausen, K. (2008). Coordinating face-to-face meetings in mobile network societies. *Information, Communication & Society, 11*(5), 640–658.

Lenz, B., & Nobis, C. (2007). The changing allocation of activities in space and time by the use of ICT – 'Fragmentation' as a new concept and empirical results. *Transportation Research Part A, 41*(2), 190–204.

Li, F., Liu, G., Liu, J., Chen, X., & Ma, X. (2016). 3D Tracking via shoe sensing. *Sensors, 16*(11), 1809. doi:10.3390/s16111809

Licoppe, C., & Smoreda, Z. (2004). Are social networks technologically embedded? How networks are changing today with changes in communication technology. *Social Networks, 27*, 317–335.

Ling, R., & Yttri, B. (2004). Hyper-coordination via mobile phones in Norway. In J. Katz & M. Aakhus (Eds.), *Perpetual contact: Mobile communication, private talk, public performance* (pp. 139–169). Cambridge: Cambridge University Press.

Lupton, D. (2016). *The quantified self.* Cambridge: Polity Press.

Mokhtarian, P. (2003). Telecommunications and travel: The case for complementarity. *Journal of Industrial Ecology, 6*(2), 43–57.

Mokhtarian, P., & Salomon, I. (2001). How derived is the demand for travel? Some conceptual and measurement considerations. *Transportation Research Part A: Policy and Practice, 35*(8), 695–719. doi:10.1016/s0965–8564(00)00013–6

Mokhtarian, P., & Varma, K. (1998). The trade-off between trips and distance traveled in analyzing the emissions impact of center-based telecommuting. *Transportation Research Part D, 3*(6), 419–428.

Ohnmacht, T., Götz, K., Schad, H., Haefeli, U., & Stettler, J. (2008). Mobility styles in leisure time: Target groups for measures towards sustainable leisure travel in Swiss agglomeration. Paper presented at the 8th Swiss Transportation Research Conference, Monte Veritá, Ascona.

Ortner, S. (1984). Theory in anthropology since the sixties. *Comparative Studies in Society and History, 26*(1), 126.

Ortner, S. (2006). *Anthropology and social theory: Culture, power, and the acting subject.* London: Duke University Press.

Pantzar, M., & Shove, E. (2010). Understanding innovation in practice: A discussion of the production and re-production of Nordic Walking. *Technology Analysis & Strategic Management, 22*(4), 447–461.

Patton, M. (1987). *How to use qualitative methods in evaluation.* Newbury Park, CA: Sage Publications.

Postill, J. (2010). Introduction: Theorising media and practice. In B. Brauchle & J. Postill (Eds.), *Theorising media and practice* (pp. 35–54). Oxford: Berghahn Books.

Rainie, L., & Wellman, B. (2012). *Networked: The new social operating system.* Cambridge, MA: Massachusetts Institute of Technology Press.

Reckwitz, A. (2002). Toward a theory of social practices: A development in cultural theorizing. *European Journal of Social Theory, 5*(2), 243–263.

Rheingold, H. (2002). *Smartmobs: The next social revolution.* New York, NY: Perseus Publishing.

Schönfelder, S., & Axhausen, K. (2010). *Urban rhythms and travel behaviour: Spatial and temporal phenomena of daily travel.* Surrey: Ashgate Publishing Ltd.

Turkle, S. (1995). *Life on the screen: Identity in the age of the internet.* New York, NY: Simon & Schuster.

Urry, J. (2007). *Mobilities.* Cambridge: Polity Press.

Vilhelmson, B., Thulin, E., & Fahlén, D. (2011). ICTs and activities on the move? People's use of time while traveling by public transportation. In S. Brunn (Ed.), *Engineering Earth: The impacts of megaengineering projects* (pp. 145–154). London: Springer.

Viry, G. (2012). Residential mobility and the spatial dispersion of personal networks: Effects on social support. *Social Networks, 34,* 59–72.

Vos, J., Derudder, B., Acker, V., & Witlox, F. (2012). Reducing car use: Changing attitudes or relocating? The influence of residential dissonance on travel behavior. *Journal of Transport Geography, 22.*

Warde, A. (2005). Consumption and theories of practice. *Journal of Consumer Culture, 5*(2), 131–153.

Webster, F. (1995). *Theories of the information society.* New York, NY: Routledge.

Wellman, B. (2002). Little boxes, glocalization, and networked individualism? In M. Tanabe, P. Besselaar, & T. Ishida (Eds.), *Digital cities II: Computational and sociological approaches* (pp. 10–25). Berlin: Springer.

Wellman, B., Quan-Haase, A., Boase, J., Chen, W., Hampton, K., Diaz, I., & Miyata, K. (2003). The social affordances of the internet for networked individualism. *Journal of Computer Mediated Communication, 8*(3). Retrieved from http://jcmc.indiana.edu/vol8/issue3/wellman.html (Accessed March 15, 2019).

Part II

The urban perspective of information and communication technology social networks and travel behaviour

8 A typology of urban analysis models

Disciplines and levels

Constantinos Antoniou, Dalit Shach-Pinsly,
Francois Sprumont and Slaven Gasparovic

Introduction

Urban analysis models have a long history. One of their characteristics is that they involve a large range of researchers from different disciplines who approach the city from different perspectives at different scales and with different objectives. However, they rarely cooperate, thus operating in distinct silos. While this may have served the community reasonably well in the past, the emergence of rich data and the speed at which transport changes nowadays, make the need for direct collaboration and interaction urgent. Urban evolution is no longer dominated by slow-moving public administration, but fast-paced private entities are increasingly becoming more active and relevant. For example transportation network companies (e.g. Uber, Lyft), dynamic vanpooling services, as well as bike- and car-sharing services are offering new services, which (have the potential to) transform the urban landscape. Furthermore, these services adapt very quickly, in response to changing regulation or traveller preferences (e.g. from single-passenger services to shared vehicles or from car-based solutions to bike- and scooter-sharing).

We first motivate the role of multiple analysis levels (focusing on these different perspectives: urban design, transportation, geography) and their relation. Naturally, each of these levels can be further distinguished into sublevels, for example in terms of level of detail, which may – in themselves – imply different approach philosophies.

For example one way to approach the different points of view could distinguish the following:

a   Planning level: aggregate representation
b   Operational level: network representation
c   Detailed analysis level: disaggregate representation

One way to exploit the interactions among these three layers follows. At the lower, detailed analysis level, it is practical to define and calculate, at the individual building level, several detailed attributes, such as visual privacy, accessibility, walkability and security. This information can then be moved up to the intermediate, operational level, where the city can be represented by, for example, a grid or collection of parcels or even compound of buildings, but without the detailed information available at the lower level. At this level, it may be practical to aggregate the detailed information of the lower level and attempt to represent entities with approximate values of the characteristics calculated at the lower level. It would thus be able to correlate network-level information with local information. At the top, the planning level, it is practical to generalize this information to yet larger areas (even across urban areas) and thus create practical models to develop further insights.

Existing models and approaches at all levels are presented in a coherent way, and a critical review is performed, aimed at highlighting the advantages and limitations of each approach. A synthesis of the needs of the different levels is also performed, providing a road map for the development of an overall architecture that can allow modellers from different disciplines and backgrounds, operating at different levels of analysis and with different objectives, to exchange views and data, with the final aim of developing a flexible and more complete model.

The problem dimensionality in space and time is then explored (e.g. 0D [point], 1D [line], 2D [level], 3D [space], 4D [space + time]), along with the issue of scale and model hierarchy. In particular, questions such as "What is scale?" and "At what scale do/can we measure each aspect?" are addressed.

Conventional analysis assumes and exploits spatial links in the modelled phenomenon. We consider social networks to provide a different landscape, in which geographical distance is not necessarily the determining factor in travel behaviour, thus linking social networks and travel behaviour in a way that moves beyond the spatial scale, and brings it more in sync with the information age. However, we also recognize that the issue is more complex and that the relation between space and social networks is more complex. Therefore, it is not presently clear that Information and Communication Technology (ICT) use has diminished the impact of distance on human interaction.

Finally, concluding remarks tie this chapter with the rest of the material in this book and provide directions for future research.

Figure 8.1 outlines the three levels that were considered (using the transportation network as a canvas). The information moves in both directions between the layers, adding value and depth to the information, and thus combines heterogeneous information in different scales.

The following paragraphs discuss the existing models and approaches at all levels.

Urban models in geography

Cities are complex systems characterized by a three-dimensional spatial structure and a strong effect of time (or past heritage) on future development. To understand activity location dynamic and urban form evolution, urban models in geography were developed. Geographers define a model as "a simplified version of reality, built in order to demonstrate certain of the properties of reality" (Haggett, 1965).

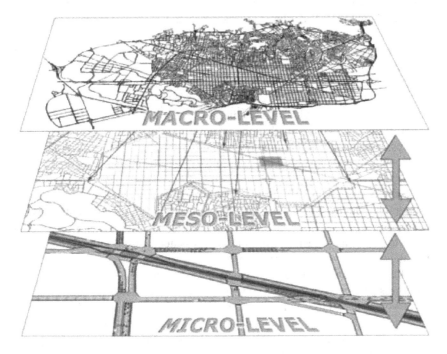

Figure 8.1 Main transportation network modelling levels (Aimsun)

The complexity inherent to urban modelling is partially due to the interest of the Geographers for the functional, morphological and social structures of spatial developments, including urban forms. Such complexity has led modellers to develop representations of urbanized areas relying on assumptions regarding the homogeneity of the land, population characteristics and preferences, among others.

For instance, to investigate the location decision of firms, Harold Hotelling (1929) opted for a one-dimension conceptualization. Hotelling's well-known location model shows, under some assumptions, that commercial activities (shops) would end up in similar central locations to maximize the consumers' numbers and their revenues. Location models – whatever their assumptions or hypotheses – are useful to understand location behavioural decisions.

Burgess's approach: the concentric zone model

Few years before Hotelling's firm location decision model, Ernest W. Burgess (1929) developed an interesting concentric zone model to describe the spatial structure and the land use development of a city – Chicago. Originally, Burgess's interest was to understand how migrants from the rural areas outside Chicago would integrate into Chicago, a fast-growing city. This concentric representation of a city describes the spatial structure of a city and its temporal evolution over time (see Figure 8.2). According to Burgess's model, the city is made of six zones:

1   Chicago's Loop – Central business district: most accessible zone where most tertiary employment is located
2   Factory zone: zone with industrial activities (ports, railyards)
3   Zone of transition: mainly contains the poorest segment of the urban population living in lowest housing conditions.
4   Working-class zone: mainly dominated by working class
5   Residential zone: higher-quality housing
6   Commuter zone: mainly high-class and expensive housing

Interestingly, already in 1929, Burgess's model assumes a relation between socio-demographic status (i.e. income) and household location settlement decision. Indeed, already in 1929, it was assumed that households with higher income could afford to commute longer distances to access better housing units further away from the central business district (CBD). Another notable feature is the consideration of time on both the expansion and reconversion of the land-use zones.

Burgess's model also presents common features with the agricultural land theory of Von Thünen (1875) or the bid rent theory of Alonso (1964). Indeed, in all these theories, land price is a function of the distance to the CBD, and all theories lead to concentric land use development. Because, for the bid rent theory, the value (or rent) of a parcel of land is related to its location, Hurd (1903) wrote, "Since value depends on economic rent and rent on location, and location on convenience, and convenience on nearness, we may eliminate the intermediate steps and say that value depends on nearness" (p. 13).

Unlike Hurd, Alonso (1964) emphasized the importance of the spatial aspect of the size of a site used in a certain city. He differentiated size and site. Alonso emphasized the importance of three elements that play a role in choosing a location. These are (1) rent, (2) transport costs and (3) size of the site. According to Alonso, the intensity of land use decreases from the city centre to the edge of the city. The CBD, that is trade, is located in the city centre, as it can afford to pay higher rent values. Leaving the CBD, industry and wholesale trade zones could be found. After them, residential areas are present, first multi-storey residential buildings and then family houses. Outside the city is a zone of agricultural land use.

In the past, a clear difference between the city and its surrounding agricultural area was evident. However, there was also a connection between them. Therefore, special attention was paid to the influence of the city on changes in the way of exploiting agricultural space. According to Vresk (1990), Von Thünen was one of the most important contributors to this topic. Von Thünen assumed that the influence of the city, which is supplied by surrounding farmlands with their agricultural products, would shape the zones of different agricultural land use around this city. Different zones of land use are due to varying location rents, which are, in turn, due to uneven transport costs of agricultural products. The first zone around the city will include intensive agricultural land use; the second zone, dedicated to woods; the third zone, to extensive agriculture. Finally, the fourth zone will be extensive cattle breeding, and the surrounding area will be wasteland.

Retrospectively, Burgess's land use concentric model suffers from several limitations (e.g. too simple, more suited to describe American cities at a time when having car was very expensive). However, Burgess model remains useful to understand urban growth in American cities during the first half of the 20th century and can be considered as a way to approach land-use understanding and modelling. As an evidence of its usefulness, researchers and academics continue to make scientific progress using the 1929 Burgess's model (interested readers might, for instance, consult the work of Meyer (2000) about "hills cities").

Hoyt's contribution: the sector model

The sector model was developed by Homer Hoyt in 1939 as a response to the shortcomings of Burgess's concentric model. An interesting feature in Hoyt's attempt to model urban structure is the combination of concentric and sectorial urban zones (Figure 8.2). Indeed, according to Hoyt, some activities do not develop in forms of rings but, instead, as sectors.

The sectors are assumed to be homogenous in terms of activity because "like attract like". For instance, heavy industry companies would benefit from positive externalities by locating themselves close to each other. The transportation system (railway and roads) is considered, and the sectors grow along important transport connections, such as railway track or major roads. More elaborate variations of the sector model include several concentric zones (for instance CBD, followed by working-class housing, middle-class housing and so on).

Hoyt also applied the sector model in the research of the social topography of a city. He found that social groups tend to concentrate within certain sectors of the city with regard to their status. Within each sector, zonal forms of concentration of social groups could be found, so their socio-economic status increases from the city centre towards the periphery. Yet, each sector should not include the whole

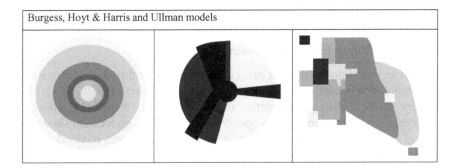

Figure 8.2 The Burgess, Hoyt and Harris and Ullman models

spectrum of social groups. Over time, the socio-economic status of all parts of the city falls as a result of the migration of certain social groups to better and newer housing away from the centre.

The multiple nuclei model from Harris and Ullman

While Hoyt's model can be considered as an improved version of Burgess's modelling effort, the multiple nuclei model proposes a different approach. Indeed, according to Harris and Ullman in their article "The Nature of Cities" (1945), developing cities do not spread evenly around a unique CBD but tend to grow around secondary centres. Such models, taking into account the existence of a main CBD and several smaller, but growing, nuclei, are a more suited representation for very large and expanding cities. Los Angeles (US) is often considered as a good example of a multiple nuclei urban development form.

Christaller and the central place theory

The concentric model, the sector model and the multiple nuclei model attempt to describe, at least partially, the urban landscape and its development, while Christaller models (1933) focus on the relations between cities. Walter Christaller, a German geographer, introduces the notion of network between human settlements – cities – and assumes that not all cities can welcome all types of activities. This idea of hierarchy within a complex network of cities explains how cities of different size can coexist and grow with their own pace.

The spatial distribution of the central places is related to their function. The main function of central places is supplying the population (consumers) of goods and services. The spatial distribution of the central places should follow a pattern of fewer larger central places that supply as many smaller central places as possible at the shortest distance. According to Christaller, central places differ in the number and types of central functions. Central places with greater number and types of central functions will have a larger gravitational area, compared to central places with lower number and types of central functions. However, the model described by Christaller also suffers from strong assumptions (evenly distributed population and resources, similar purchasing power, perfect competition, etc.).

Urban modelling: a myriad of approaches

Of course, in addition to the preceding classical models, other models also study the spatial structure of a city. From the research of Sjoberg (1960), the *evolutionary model* of the spatial structure of the city was developed. According to Sjoberg, a city is the product of a specific society in a given location (in which part of the world) and temporal evolution. Three different stages of city development are distinguished even though the evolution process is continuous: (1) the pre-industrial city, (2) the industrial city and (3) the metropolis. The pre-industrial city was described as a city with its walls, with no particular functional structure, with a large density of population falling sharply towards the edge of the city. The industrial city is a functional zoning with different industrial and residential zones, and the difference in population density between the city centre and the edge of the city is not as pronounced as in the pre-industrial city. In the final stage, the metropolis is being developed at the time of tertiary and quaternary processes of society, with distinct functional zoning and a smaller population density in the city centre due to the CBD. The basis of all the aforementioned models is the interdependence of people and the social environment, that is human ecology.

While urban models have a long history, recent progress (both conceptual and computational) has led researchers to consider cities or regions as complex systems (see White, Engelen, & Uljee, 2015). Land Use and Transport Integrated (LUTI) models, which became popular in the 1960s thanks to Lowry (1964), take into consideration the interactions between transport infrastructure and land use. While most of the previous model (Burgess, Hoyt & Harris and Ullman models) were descriptive, LUTI models allow the forecasting of the effect of land-use modification on the transport system and vice versa (Badoe & Miller, 2000). UrbanSim from Paul Waddell (2002) and Ilute from Erik Miller (Salvini & Miller, 2005) are examples of LUTI models allowing the development of advanced scenarios associating land-use development and the transport system.

Geographers also use discrete-choice models to analyze urban systems (Hutchinson, Nijkamp, & Batty, 2012) and residential choice strategies. Interestingly LUTI, discrete choice models and agent-based models are also commonly used by transport engineers and academics. This suggests that both transport researchers/practitioners and geographers use, to a certain extent, similar methodologies to approach different research questions.

Finally, it is also worth distinguishing descriptive urban models from analytical ones. Indeed, rather obviously, the Burgess, Hoyt and Harris and Ullman models are mainly descriptive, while, for instance, LUTI models are considered as analytical. However, it would be naive to summarily consider older urban models as descriptive and recent ones as analytical. The agricultural land theory of Von Thünen (1875) constitutes a good example of an analytical model being developed in the late 19th century. On the other hand, models from Burgess, Hoyt and Harris and Ullman are often seen as the ancestors of urban models by geographers and will continue to be studied by thousands of students worldwide because they allow the introduction of key concepts, such as spatial segregation, urban sprawl and transport corridors.

Urban models in transportation networks analysis

Mobility is a key aspect and source of trouble in every urban area. While never a means in itself, mobility is a key element of quality of life. There are two main components of transportation network analysis: the demand and the supply sides. The supply side is a bit easier to grasp, as it corresponds to the physical network representation. As such, it is somewhat easier to understand what it entails. Many of its properties can be measured, and there are readily available representations (e.g. OpenStreetMaps). The demand side, on the other hand, is not observed, and – in fact – arguably not observable with current technology. This includes the exact demand patterns that are being loaded onto the transportation network, resulting in traffic conditions.

One characteristic of transportation network analysis is that, while the scale can naturally change, the models are typically rather closely tied to the physical representation of the network. While traditionally transportation network analysis has been focusing on the roads, and all the passenger vehicles that load them, it has been increasingly recognized that the transportation network comprises many other interacting elements, in fact not only the pedestrian and cycling movements but also freight services and vehicles, and public transport (on- and off-road). Furthermore, new modes and services are becoming increasingly available (e.g. transportation network companies, such as Uber and Lyft; vehicle-sharing services; and soon automated and autonomous vehicles). Transportation network analysis tools need to keep evolving in order to capture these developments.

One interesting property of transportation networks is that they are not static, but rather dynamic. This dynamic nature can be manifested in different ways and at different scales. For example, in a short-term scale, a widespread snowstorm can affect the entire network pretty much uniformly, while a local incident would only have a concentrated impact (which, of course, can take bigger dimensions, depending on the prevailing conditions). Other changes, more long term, involve for example the change in the direction of a one-way street or the addition of new facilities, for example the creation of a pedestrian zone or the repurposing of a traffic lane as a bus lane.

Transportation network representation is not uniquely defined. Instead, depending on the scale of the application, the form of the transportation network that is used to represent the underlying region can change; see for example Figure 8.1. If we are interested in modeling mobility in a wider region, for example, we might choose a representation that is based on the main corridors, leaving some of the minor local streets out. If, on the other hand, the point is to model local impacts on the CBD of for example urban delivery restrictions, then it is important that all facilities are fully modelled.

These differences in the scale do not imply that one model is superior to the other. For example, using a very detailed model to capture overall trends in a wider region might not simply result in a "waste" of resources, but it could actually provide suboptimal results. This is due to the fact that in developing a more detailed model than what is suitable for the task at hand, we are introducing potential errors (through the extra aspects that we are modeling). The remainder of this section provides an overview of the development of transportation network models.

Travel demand models/planning applications

Transportation network analysis has essentially started in the middle of the previous century (post-war era), with applications of the four-step model. The first comprehensive application of the four-step model has been in the Chicago Area Transportation System (CATS), which provided a validation of the concept in a large-scale. Of course, there were limitations associated with that application, e.g. the fact that it was heavily focused on highway development. While rather aggregate in nature, this model was still computationally very demanding (for that era's "computing" capabilities) and required a long time to be developed. The objective of such models is typically to evaluate the impacts of policies and developments, prior to their implementations, so that they can be optimally planned. An important aspect of this task is being able to assess the overall impacts at the network level. In order to estimate these impacts, one needs to consider a large number of different interacting forces, which are made possible through such representations.

Still, in the following decades, a number of refinements and extensions were introduced, aiming at improving the ability of the model to respond to various changes, and also increase its level of detail. These refinements led to the development of the activity-based approach. This reflects the movement from aggregate models to disaggregate models. This allows the modeling of individual traveller characteristics and thus added realism through factors, such as heterogeneity and response to information.

Transportation system simulation/operations models

The transportation network is a very complex system of heterogeneous facilities, for example roads, sidewalks, bike paths, pedestrian zones and shared spaces, as well as dedicated facilities, such as metro lines.

In order to meet the broad needs for traffic simulation (Antoniou, 2014), a wide range of traffic simulation models with different properties and characteristics have been developed over the last several decades. Traffic simulation models can be classified according to multiple dimensions, such as level of granularity or fidelity with which they model traffic. Choosing the appropriate model is a fundamental decision for the success of any simulation project. Often, modellers tend to know a tool well and use it for all tasks, irrespective of its suitability. Ideally, one should aim to use the right (level of) tool for each task. Traditionally, transportation network analysis models have been categorized into three levels, depending on the level of detail (also seen in Figure 8.1). For an overview of traffic simulation, with many examples of state-of-the-art models, see Barcelo (2010).

Microscopic traffic simulators (e.g. AIMSUN [Casas, Ferrer, Garcia, Perarnau, & Torday, 2010], CORSIM [Halati, Lieu, & Walker, 1997], DRACULA [Liu, 2010], MITSIMLab [Ben-Akiva et al., 2010], Paramics [Sykes, 2010], Trans-Modeler [Caliper, 2018], VISSIM [Fellendorf & Vortisch, 2010]) entail detailed models of driver behaviour, comprising car-following, gap-acceptance, lane-changing and other disaggregate behavioural models. These models operate at the individual driver level and utilize a large number of parameters. The internal clock of such models often operates at the level of a fraction of a second, meaning that decisions about the behaviour of the travellers and the movement of the vehicles may take place several times per second. Microscopic traffic simulation allows for a much higher resolution and detail in modeling (than mesoscopic and macroscopic models), and as such, it may provide more accurate results. Naturally, this level of detail comes at a computational cost. In earlier years, this meant that microscopic traffic simulation could only be used for limited networks. However, recent advances in computing (more important, not only hardware developments but also parallel computing and more efficient algorithms) have made the use of microscopic simulation practical even for real-time or near-real-time applications.

Lower resolution (macroscopic or mesoscopic) simulation models usually represent traffic dynamics in a more aggregate way and may update the vehicle position and characteristics every few seconds. Macroscopic models are well established and mature, as they are being used for several decades already. Computationally they offer considerable advantages, and therefore, they can be used to evaluate large numbers of scenarios or very large networks. Examples of macroscopic models include SATURN (Hall & Willumsen, 1980) and METANET (Messmer & Papageorgiou, 1990).

Mesoscopic traffic simulation models represent individual vehicles and advance along links consisting of a "moving" part and a "queuing" part of each link. Vehicles move in the moving part according to a speed, which depends on the prevailing conditions, usually density. Vehicles that are queuing at the downstream end of the link are advanced based on the output capacity of the segment.

Microscopic and macroscopic traffic models had monopolized the traffic simulation scene for a couple of decades until the early 1990s, when mesoscopic simulation came to prominence. The idea of mesoscopic simulation is that each vehicle is still simulated independently, but – whenever possible – at a lower level of detail. For example, when vehicles are moving in an uncongested freeway, it is not necessary to track the individual lane that they are following. However, when they reach a signalized intersection, they need to be assigned to lanes so that for example vehicles making a right turn do not queue and block vehicles moving straight. Emerging Traffic Estimation and Prediction (or Dynamic Traffic Assignment, DTA) systems, used in real-time applications for traffic estimation and prediction, are simulation-based and use mesoscopic simulation models to capture traffic dynamics (e.g. academic systems such as DynaMIT [Ben-Akiva et al., 2010; Ben-Akiva, Koutsopoulos, Antoniou, & Balakrishna, 2010], DYNAS-MART [Mahmassani, 2001] and DynusT [Chiu, Nava, Zheng, & Bustillos, 2011], as well as commercial such as Dynameq [Mahut & Florian, 2010]).

Table 8.1 succinctly presents some of the comparative advantages and disadvantages of each of the previously mentioned types of models. It is important to recognize that there is no dominant model or family of models. Depending on the application (scope, area, resource availability, data availability), different models can be more suitable.

Current trends/changes in data availability

For the majority of transportation network analysis modeling (i.e. in the second half of the 20th century) one of the biggest challenges has been the (un)availability

Table 8.1 Advantages and disadvantages of classes of traffic simulation models

	Advantages	*Disadvantages*
Microscopic	• More detailed/accurate than mesoscopic/macroscopic • Ability to explicitly model details such as traffic signal strategies, road geometry • Modern micro simulators have graduated from turning-fraction type approaches to OD/path-based modelling	• Computationally expensive • More and more detailed data required • Longer time to develop, calibrate and validate model • Calibration effort is very demanding • Microscopic detail may not be necessary for online applications and contingency planning. • Even though micro simulators are getting faster, mesoscopic models would always allow larger networks to be simulated, more scenarios to be modelled, longer prediction horizons
Mesoscopic	• Practical compromise between Micro and Macro • Combination of aggregate and disaggregate models • Actively researched, especially in the context of real-time forecasts and generation of contingency plans • Allows for longer prediction horizons than microscopic simulation (up to 1–2 hours)	• Not many practical, large-scale, online applications completed (however, many research/ evaluation studies have been undertaken, and several are underway)
Macroscopic	• Computationally very efficient • Mature, widely deployed technology • Can be used to quickly evaluate a large number of contingency plans, thus filtering available alternatives to a handful that can then be evaluated in more detail (by another system)	• Dated technology • Not very extensible without moving to mesoscopic/ microscopic solutions • Mostly limited in practice to planning/off-line applications

of data. Modellers were forced to develop travel demand models using a few thousands of survey responses and developers of operational models often only had data from a couple hundred loop detectors in an entire city or region. This is rapidly changing, with data being increasingly available (e.g. cell phone signals from telecommunication operators, social media data, records of Bluetooth detections, counts from cameras), which provide richer representations of the dynamics of the travellers. The abundance of data has – in some cases – changed the modelling approach from trying to maximize the value of each individual data point to trying to extract meaningful structure from the wealth of available data.

Data-driven techniques are thus becoming increasingly available (e.g. Antoniou, Koutsopoulos, & Yannis, 2013; Papathanasopoulou & Antoniou, 2015), which are expected to play a significant role in the future. Furthermore, the role of big data and social network data (e.g. Chaniotakis, Antoniou, & Pereira, 2016; Chaniotakis, Antoniou, Aifadopoulou, & Dimitriou, 2017) in supporting transportation modelling applications has been long identified, and related applications are gradually maturing and producing tangible results.

Urban models in architecture

Urban models in architecture and urban design were developed over the years in several modes, mainly for understanding the space and place of the city at a more detailed level. There is not one common definition of what an urban model in urban design is, however, they can be determined by time and technology. In addition, these models can be specified by three main categories and can be described as follows: (1) city mapping, (2) urban planning models and (3) urban design analysis. Following is a short description with examples of these three categories.

City mapping

The first-known attempt for analysing and modelling the city was done through mapping and was made by the Nolli map (Nolli, 2019). The Nolli map was designed by Giambattista Nolli in 1748 for measuring density in cities. He decided to represent the city space by representing the footprints of the buildings in black and the open public spaces in white, including enclosed public spaces of churches, palaces and monasteries. This representation method was widely accepted in the urban design field as a way for representing open public spaces. In his new development, Nolli made several important innovations for the urban field: (1) re-orienting the city map from east to the magnetic north for better capturing the city topography, (2) understanding the city through a black/white mapping layout of the buildings' footprints and (3) the accuracy of the map was significant for that time. The Nolli map served as a basis for all Roman mapping and planning until the 1970s (Hwang & Koile, 2005).

Urban planning models

Over the years cities, and years after Nolli's map, urban models were developed aiming at integrating the city life with physical structure based on the spirit of the times, for example the democratic spirit which affected urban development in Greece (338–1200 BCE), the impact of Islam on urban cities design (7th century AD), and the Renaissance and Baroque model cities that were inspired by the new urban lifestyle in Europe, which advocated the separation between religion, politics, government and commerce (15th–18th century AD; Braudel, 1982).

One of the turning points in the modern age of urban model development was the concept of the Garden City movement that was initiated in 1898 by Sir Ebenezer Howard (Howard, 1902; Clark, 2003). The Garden City approach was developed as a response to the problems suffered by British cities following the Industrial Revolution, which was characterized by density, pollution, poor sanitary conditions, gaps and social tensions. This concept was also visualized by a set of drawings showing the layout of the model. Garden Cities were intended to be planned, containing proportional areas of residences, industry and agriculture while housing 32,000 people on a site of 6,000 acres, planned based on a concentric pattern with open spaces, public parks and six radial boulevards, extending from the centre. Howard visualized a cluster of several Garden Cities as satellites of a large central city of 250,000 people linked by roads and a rail (Goodall, 1987; March, 2004).

Modern times opened the stage for urban models that capture the city development in one clear concept and served as a planning tool for the city in all levels and opened the door to the development of the modern movement

The modern movement ·

In the first half of the 20th century, the term *modern architecture* emerged from revolutions in technology and building materials and influenced urban planning mainly after World War II. The modern architecture movement was based on new construction technologies; the use of reinforced concrete, steel, glass; and the industrialization of constructed building details and material development. This revolution inspired designers to break away from the neoclassical and "old" development and invent new concepts for design (Frampton & Futagawa, 1983).

One of the modern movement pioneers was Le Corbusier, an architect, designer and urban planner. He had an outstanding contribution to the modern movement by learning to take advantage of using reinforced concrete in buildings and defined five basic principles of planning stemming from the use of reinforced concrete: (1) pilotis, lifting the structure off the ground, (2) building a skeleton based on columns to form its internal geometric division, (3) long strips of ribbon windows to allow open views, (4) open space for living and (5) a roof garden. In addition, in 1922, Le-Corbusier presented a design of a city for 3 million people "Ville Contemporaine". This new concept of urban planning was seeking a city whose residents would live and work in a group of identical 60-storey apartment buildings surrounded by lower identical apartment blocks and a large open park. The houses were modular, aiming to be mass-produced and assembled into apartment blocks, neighbourhoods and cities (Le Corbusier, 1923; Tietz, 1999).

In response to the modern movement approach that promoted zoning and private motor vehicles and led to neglecting city centres, the New Urbanism movement emerged.

New Urbanism

The New Urbanism movement, developed during the 1980s in the United States, advocates design-based strategies based on "traditional" urban forms to help arrest suburban sprawl, inner-city decline, build and rebuild neighbourhoods, towns and cities (Bohl, 2000). Its objectives include the restoration of urban centres and towns within coherent metropolitan regions, the reconfiguration of sprawling suburbs into communities of "real" neighbourhoods, the conservation of natural ecosystems and biodiversity and the preservation of historic and culturally significant architecture, neighbourhoods and landscapes. The variable scales range from the individual building to entire cities and regions. In the regional context, New Urbanism advocates a full range of urban settings, including rural hamlets, villages, small towns, dense urban neighbourhoods and districts, to provide compact development alternatives appropriate to each setting (Talen, 1999).

The New Urbanism principles are demonstrated in a development plan for the rural community of Tsawwassen, British Columbia, Canada, a bedroom community in which 75% of its residents commute to Vancouver (Roehr & Kunigk, 2009). The Southlands project attempts to integrate local food production with urban living, eliminating the boundaries between residential and agricultural spaces, and facilitates opportunities for each resident to farm and produce food. Agriculture is also integrated at a social and cultural level. The town centre serves as a market place for produce, retail, restaurants and civic buildings, including a market space for local food producers to sell their wares. Walkability remains a priority, and the town centre is a 10-minute walk from all neighbourhood homes. Southlands' residential buildings are diverse, including single to multiple-family homes and units for senior citizens and students.

Landscape Urbanism developed in response to criticisms of new urbanism, aiming at organizing cities through the design of the landscape rather than the design of its buildings (Gray, 2006). It offers a way to consider a complex urban condition and reciprocal implications of the city in the landscape on one another (Mostafavi & Najle, 2003). The High Line project in New York City was inspired by the Landscape Urbanism, integrating green-ecological based concepts into the city.

These urban models were mainly based on current site analysis and continue reaction to existing earlier urban models and theories. However, the emergence of the technology push has changed this approach.

Urban design analysis

In the last decade the focus of urban design analysis shifted from analyzing the actual footprints layout and building volumes towards analyzing qualitative and quantitative aspects of the environment that have a significant influence on the built environment. This shift was a result of major technological changes and advancements that influence the way we understand the urban environment.

This technology push enables the measurement of qualitative aspects of the built environment in quantitative terms and allows a better understanding of the quality of the built environment, as can be obtained between buildings. Examples of such qualities are *walkability* (Moudon, 2006), which influences the way we walk through the built environment; *visibility* (Fisher-Gewirtzman, Pinsly, Wagner, & Burt, 2005), how far one can observe the built environment; level of *personal security* (Shach-Pinsly, 2018), feeling safe and secured in open public spaces; *way-finding* analysis; and *privacy* in urban areas, among others. This new form of describing qualitative aspects influences the way we use and develop the built environment; therefore, the shift of urban analysis requires new approaches for using technology and data-base sources to analyse the urban environment.

Hence, current urban design analysis focuses on developing qualitative and quantitative methods for measuring and analysing urban spatial issues and various environmental aspects that affect the built environment. The analysis can be visualized using various models. Figure 8.3 demonstrates two models of measuring different aspects of the environment: (1) measuring walkability by using a surface mapping method and (2) measuring privacy ratio in a neighbourhood.

The "Neighborhood Walkability Score" (Urban Form Lab-UFL, University of Washington) in Figure 8.3 (left) was designed as a tool for modelling the walkability aspects of physical environments in order to promote walking and physical activities in a neighbourhood scale. Using the tool assists in understanding urban spaces and supports urban development. Measuring visibility and privacy, Figure 8.3 (right) assists in understanding our close environment. New technologies

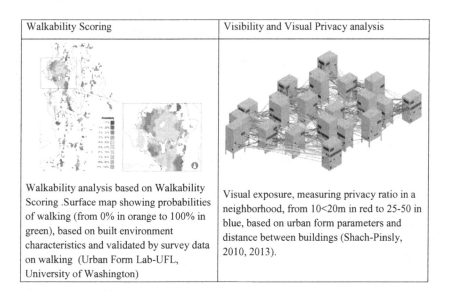

Walkability Scoring	Visibility and Visual Privacy analysis
Walkability analysis based on Walkability Scoring .Surface map showing probabilities of walking (from 0% in orange to 100% in green), based on built environment characteristics and validated by survey data on walking (Urban Form Lab-UFL, University of Washington)	Visual exposure, measuring privacy ratio in a neighborhood, from 10<20m in red to 25-50 in blue, based on urban form parameters and distance between buildings (Shach-Pinsly, 2010, 2013).

Figure 8.3 Urban design analysis, walkability analysis based on Walkability Scoring (Urban Form Lab-UFL, University of Washington, http://depts.washington. edu/ufl/; Lee and Moudon, 2006; Berke, Koepsell, Moudon, & Hoskins, 2007) and visual exposure analysis (Shach-Pinsly, 2010, 2013)

also enable us to understand our close environment, whether the buildings are too close to each other, invade our privacy or have enough distance between buildings to allow open visibility to the view.

To conclude, urban design analysis is using modern technology to slowly expand its scale of analysis and connect to wider scales (transport/middle level; geography/top level) without giving up the small scale of the building. This allows for developing a more comprehensive analysis for wider urban areas.

Conclusion: integrating disciplines, levels and urban models for better implementation in the built environment

As discussed in this chapter, each discipline design and implement its own urban models for better develop and manage the built environment. However, currently, the technology push and the emergence of a wide variety of sources of open, ubiquitous data have the potential to encourage and support specialists from different disciplines to cooperate. This cooperation and interaction will allow them to analyse this integration and to look for the connection points that will suit all levels. This analysis is challenging, having many advantages and limitations of connecting/relating the differences in scales. Therefore, when thinking towards the development of future models, we need to take into consideration different scales that will be able to integrate detailed parameters (from the lower level, urban design) with wider metropolitan parameters (the higher level, geography) for better the connecting between these two scales that will be transformed into the middle scale (the transportation level) to develop tailored transportation network models for a better functionality of the built environment.

In the lower level of urban design, additional quantitative models need to be developed. These models need to aspire for data-driven techniques, defining and locating specific urban details and variables that characterize specific qualities of the built environment and can be measured in quantitative terms, such as distances between buildings, distances between a building and the fence surrounds the building courtyard or distance between the fence and the sidewalk edge. These urban details and variables can be integrated into the middle-level models of transportation, where the detailed environment can be taken into consideration and be related to the higher level of geography by understanding the built environment as a whole and complex environment.

This requires a paradigm shift in model development, which currently looks narrowly at the parameters that seem relevant only at the eye level of each modeller. We need to leave our comfort zones and embrace new ideas from the other disciplines. This is not going to be easy and will certainly be faced with resistance and inertia. Before being able to develop a new generation of theory-based models, which combine and integrate scientific disciplines, we can first explore data-driven models, which can provide a tangible demonstration of the benefits of "joining forces". Armed with such information, we will then be able to push forward to a new era of theory-based (or consistent) data-driven models, which are expected to provide vastly superior results.

This may need a change in the way we look at the built environment, promote new model development and use new data-driven techniques that are increasingly available. In addition, the three different layers may inspire different working-approaches to be implemented in the analysis, thus increasing the challenge of the integration. Therefore, specialists of the different levels need to take into consideration these gaps of knowledge and analysis techniques for enabling this integration of the three levels.

References

Alonso, W. (1964). *Location and land use: Toward a general theory of land rent*. Cambridge, MA: Harvard University Press.

Antoniou, C. (2014). Traffic simulation. In M. Garrett (Ed.), *Encyclopedia of transportation. Social science and policy*. Thousand Oaks, CA: Sage Publications.

Antoniou, C., Koutsopoulos, H. N., & Yannis, G. (2013). Dynamic data-driven local traffic state estimation and prediction. *Transportation Research Part C: Emerging Technologies, 34*, 89–107.

Badoe, D. A., & Miller, E. J. (2000). Transportation – land-use interaction: Empirical findings in North America, and their implications for modeling. *Transportation Research Part D: Transport and Environment, 5*(4), 235–263.

Barcelo, J. (Ed.). (2010). *Fundamentals of traffic simulation*. New York, NY: Springer.

Ben-Akiva, M., Koutsopoulos, H. N., Antoniou, C., & Balakrishna, R. (2010). Traffic simulation with DynaMIT. In J. Barcelo (Ed.), *Fundamentals of traffic simulation* (pp. 363–398). ISBN: 978-1-4419-6141-9, New York, NY: Springer.

Ben-Akiva, M., Koutsopoulos, H. N., Toledo, T., Yang, Q., Choudhury, C. F., Antoniou, C., & Balakrishna, R. (2010). Traffic simulation with MITSIMLab. In J. Barcelo (Ed.), *Fundamentals of traffic simulation* (pp. 233–268). ISBN: 978-1-4419-6141-9, New York, NY: Springer.

Berke, E., Koepsell, T., Moudon, A., & Hoskins, R. (2007). Association of the built environment with physical activity and obesity in older persons. *American Journal of Public Health, 97*(3), 1–7.

Bohl, C. C. (2000). New urbanism and the city: Potential applications and implications for distressed inner-city neighborhoods. *Housing Policy Debate, 11*(4), 761–801.

Braudel, F. (1982). *Civilization and capitalism, 15th-18th century, Vol. III: The perspective of the world*. Berkeley, CA: University of California Press.

Burgess, E. W. (1929). Urban areas. In T. V. Smith & L. D. White (Eds.), *Chicago: An experiment in social science research* (pp. 113–138). Chicago, IL: University of Chicago Press.

Caliper Corporation. (2018). *TransModeler Traffic Simulation software – version 4.0 User's guide*. Caliper Corporation.

Casas, J., Ferrer, J. L., Garcia, D., Perarnau, J., & Torday, A. (2010). Traffic simulation with aimsun. In *Fundamentals of traffic simulation* (pp. 173–232). New York, NY: Springer.

Chaniotakis, E., Antoniou, C., Aifadopoulou, G., & Dimitriou, L. (2017). Inferring activities from social media data. *Transportation Research Record: Journal of the Transportation Research Board, 2666* (1), 29–37.

Chaniotakis, E., Antoniou, C., & Pereira, F. (2016). Mapping social media for transportation studies. *IEEE Intelligent Systems, 31*(6), 64–70.

Chiu, Y. C., Nava, E., Zheng, H., & Bustillos, B. (2011). *DynusT user's manual.* Tucson, AZ: University of Arizona.

Christaller, W. (1933). *Die Zentralen Orte in Süddeutschland: Gustav Fischer, Jena.* English translation, 1966: Central Places in Southern Germany. Englewood Cliffs: Prentice Hall.

Clark, B. (2003). Ebenezer Howard and the marriage of town and country. *Archives of Organizational and Environmental Literature, 16*(1), 87–97.

Fellendorf, M., & Vortisch, P. (2010). Microscopic traffic flow simulator VISSIM. In *Fundamentals of traffic simulation* (pp. 63–93). New York, NY: Springer.

Fisher-Gewirtzman, D., Pinsly, D. S., Wagner, I. A., & Burt, M. (2005). View-oriented three-dimensional visual analysis models for the urban environment. *Urban Design International, 10*(1), 23–37.

Frampton, K., & Futagawa, Y. (1983). *Modern architecture 1851–1945.* New York: Rizzoli.

Goodall, B. (1987). *Dictionary of human geography.* London: Penguin.

Gray, C. D. (2006). *From emergence to divergence: Modes of landscape urbanism.* MArch thesis. Edinburgh, UK: School of Architecture, University of Edinburgh.

Haggett P. (1965). *Locational analysis in human geography.* Londres: E. Arnold.

Halati, A., Lieu, H., & Walker, S. (1997). CORSIM-corridor traffic simulation model. In *Traffic congestion and traffic safety in the 21st century: Challenges, innovations, and opportunities.* Urban Transportation Division, ASCE; Highway Division, ASCE; Federal Highway Administration, USDOT; and National Highway Traffic Safety Administration, USDOT. New York: American Society of Civil Engineers.

Hall, M., & Willumsen, L. G. (1980). SATURN-a simulation-assignment model for the evaluation of traffic management schemes. *Traffic Engineering & Control, 21*(4).

Harris, C. D., & Ullman, E. L. (1945). The nature of cities. *The Annals of the American Academy of Political and Social Science, 242*(1), 7–17.

Hotelling, H. (1929). Stability in competition. *The Economic Journal, 39*(153), 41–57.

Howard, E. (1902). *Garden cities of tomorrow.* London: S. Sonnenschein & Co. Ltd. (at Google Books).

Hoyt, H. (1939). *The structure and growth of residential neighborhoods in American cities.* Washington, DC: US Federal Housing Administration.

Hurd, R. M. (1903). *Principles of city land values.* New York, NY: The Record and Guide.

Hutchinson, B. G., Nijkamp, P., & Batty, M. (Eds.). (2012). Optimization and discrete choice in urban systems. *Proceedings of the International Symposium on New Directions in Urban Systems Modelling* held at the University of Waterloo, Canada, July 1983 (Vol. 247). Springer Science & Business Media.

Hwang, J. E., & Koile, K. (2005). Heuristic Nolli map: A preliminary study in representing the public domain in urban space. In *Proceedings of 9th International Conference Computers in Urban Planning and Urban Management.* London, UK.

Le Corbusier. (1923). *Vers une architecture* (pp. XVIII-XIX). Flammarion edition (1995). Flammarion publication. Paris, France.

Lee, C., & Moudon, A. V. (2006). Environmental correlates of walking for transportation or recreation purposes. *Journal of Physical Activity and Health, 3*(Supplement 1), S77–98.

Liu, R. (2010). Traffic simulation with DRACULA. In *Fundamentals of traffic simulation* (pp. 295–322). New York, NY: Springer.

Lowry, I. S. (1964). *A model of metropolis* (No. RM-40535-RC). Santa Monica, CA: Rand Corporation.

Mahmassani, H. S. (2001). Dynamic network traffic assignment and simulation methodology for advanced system management applications. *Networks and Spatial Economics*, *1*(3–4), 267–292.

Mahut, M., & Florian, M. (2010). Traffic simulation with Dynameq. In *Fundamentals of traffic simulation* (pp. 323–361). New York, NY: Springer.

March, A. (2004). Democratic dilemmas, planning and Ebenezer Howard's Garden City. *Planning Perspectives*, *19*, 409–433.

Messmer, A., & Papageorgiou, M. (1990). METANET: A macroscopic simulation program for motorway networks. *Traffic Engineering & Control*, *31*(9).

Meyer, W. B. (2000). The other Burgess model. *Urban Geography*, *21*(3), 261–270.

Mostafavi, M., & Najle, C. (Eds.). (2003). *Landscape urbanism: A manual for the machinic landscape*. London: Architectural Association.

Moudon, A. V., Lee, C., Cheadle, A. D., Garvin, C., Johnson, D., Schmid, T. L. . . . Lin, L. (2006). Operational definitions of walkable neighborhood: Theoretical and empirical insights. *Journal of Physical Activity and Health*, *3*(S1), S99–S117.

The Nolli Map website. (2019). Nolli Map of Rome. Retrieved from http://nolli.uoregon. edu/

Papathanasopoulou, V., & Antoniou, C. (2015). Towards data-driven car-following models. *Transportation Research Part C*, *55*, 496–509.

Roehr, D., & Kunigk, I. (2009). Metro Vancouver: Designing for urban food production. *Berkeley Planning Journal*, *22*(1).

Salvini, P., & Miller, E. J. (2005). Ilute: An operational prototype of a comprehensive microsimulation model of urban systems. *Networks and Spatial Economics*, *5*, 217–234.

Shach-Pinsly, D. (2010). Visual exposure and visual openness analysis model used as evaluation tool during the urban design development process. *Journal of Urbanism (JU)*, *3*(2), 161–184.

Shach-Pinsly, D. (2013). From qualitative to quantitative: A conceptual framework for transforming qualitative aspects of environmental quality into quantitative terms for the benefit of the designers' work. *International Researchers Journal*, *2*(1), 174–183.

Shach-Pinsly, D. (2018). Measuring security in the built environment: Evaluating urban vulnerability in a human-scale urban form. In Special Issue 'Measuring human-scale urban form and its performance': *Landscape and Urban Planning*.

Sjoberg, G. (1960). *The preindustrial city, past and present*. New York, NY: The Free Press.

Sykes, P. (2010). Traffic simulation with paramics. In *Fundamentals of traffic simulation* (pp. 131–171). New York, NY: Springer.

Talen, E. (1999). Sense of community and neighbourhood form: An assessment of the social doctrine of new urbanism. *Urban Studies*, *36*(8), 1361–1379.

Tietz, J. (1999). *The story of architecture of the 20th century*. Publisher: Konemann.

Von Thünen, J. H. (1875). *Der isolirte staat in beziehung auf landwirtschaft und nationalökonomie* (Vol. 1). Berlin: Wiegandt, Hempel & Parey.

Vresk, M. (1990). *An introduction to urban geography* (In Croatian: *Osnove urbane geografije*). Zagreb: Školska knjiga.

Waddell, P. (2002). UrbanSim: Modeling urban development for land use, transportation and environmental planning. *Journal of the American Planning Association*, *68*(3), 297–314. (preprint)

Walkability scoring. (2019). Urban Form Lab-UFL, University of Washington. Retrieved from http://depts.washington.edu/ufl/

White, R., Engelen, G., & Uljee, I. (2015). *Modeling cities and regions as complex systems: From theory to planning applications.* Cambridge, MA: Massachusetts Institute of Technology Press.

9 Travel behaviour and social network interactions with the urban environment, a review

João de Abreu e Silva, Giannis Adamos,
Domokos Esztergar-Kiss, Jasna Mariotti,
Maria Tsami and Mario Cools

Introduction

There is an increasing interest among researchers on how social interactions affect travel decisions and activity location choices (e.g. Wang & Lin, 2013; van den Berg, Kemperman, & Timmermans, 2014; Sadri, Lee, & Ukkusuri, 2015). An important research objective is to gain a more profound understanding of the specific characteristics of user interactions, which are partially generated and reported on social networks (Carrasco, Hogan, Wellman, & Miller, 2008a).

Social networks and urban environment characteristics have been considered as relevant elements that shape travel behaviour (Axhausen, 2008). Research on the interactions between travel behaviour and land use is relatively abundant. Furthermore, research on the interactions between travel behaviour and social networks emerged recently. In contrast, the existence of studies combining both is much scarcer. Therefore, the main objective of this chapter is to review theoretical and empirical studies that looked at the influences of social networks and the urban environment. More specifically, we look at literature that provides evidence regarding the constraints and opportunities posed by social networks and the relationship with the urban environment. After all, the activity-based approach and the use of behavioural theories from social psychology (e.g. Azjen's theory of planned behaviour) include in their considerations both social and spatial interactions that could act as restrictions, incentives or even substitutes of travel. Social networks could impose constraints, create opportunities and contribute to the creation of behavioural norms and materialize peer pressure. The urban environment characteristics could create difficulties or facilitate the emergence of face-to-face social interactions. But different types of urban environments and morphology could also increase social interactions and contribute to the emergence of social networks with different characteristics. Technology and particular Information and Communication Technologies (ICTs) could mediate the relationships between social networks and the urban environment (Axhausen, 2007; van den Berg, Arentze, & Timmermans, 2013).

The contributions of the review are twofold. First, the inclusion of key works that are scattered across different research domains provides a transversal perspective on this topic. Second, a framework of the relations between travel behaviour, social networks and built environment is built, based both on previous theoretical contributions and on empirical results.

The remainder of this chapter is structured as follows: First, the effects of social networks on travel behaviour are discussed. Consequently, the influence of the urban environment on social networks is elaborated. The third section discusses the influence of social networks on different appropriations and uses of the urban space. Finally, the main conclusions of the reviewer are depicted.

The influence of social networks on travel behaviour

The study of social networks and its impacts on mobility and interrelations with the urban environment has gained attention in recent decades. A very relevant set of human activities requires contact and social interactions (Páez & Scott, 2007). The study of social networks has antecedents that come from the decades of the 1970s and 1980s. In particular, Hägerstrand (1970) pointed out the coupling restrictions related to the common activities with other individuals within the scope of spatio-temporal geography, Pas (1985) highlighted interpersonal constraints in the research of space–time–activities models, and Salomon (1985) indicated the sense of belonging as one of the motivators of mobility. However, the relationships between the study of mobility and consumer theory, which underlies a concept of individualism, which has methodological implications (Páez & Scott, 2007), have relegated the role of social influences to the background. Behaviour is seen as the result of individual decisions (individual agency), implying an indifference to social influences that may include the actions and preferences of others. However, individual behaviour is also explained by social interactions between individuals belonging to the same social network (Carrasco, Hogan, Wellman, & Miller, 2008b). The growing awareness of the importance of social interactions in mobility seems to be associated with a number of aspects:

- Theories of social psychology, in particular Ajzen's theory of planned behaviour, which explicitly considers peer pressure (Páez & Scott, 2007), as evidenced by subjective norms
- The weight of social movements, with the complexity resulting from the synchronization of individuals' time budgets (Harvey & Taylor, 2000). In particular, temporal budgets associated with leisure have grown, and the boundaries between work and leisure/social activities have become more porous (Carrasco et al., 2008b).
- Telecommunications, ICT and social media, since these enable the connection between the members of a social network between face-to-face contacts (Carrasco et al., 2008a)

The study of social interactions is strongly linked to the characterization of social networks, and the influence of their characteristics on mobility (e.g. Carrasco et al., 2008a; Lin & Wang, 2014; Axhausen, 2008; Arentze & Timmermans, 2008). Social networks can be sources of information spill-overs, facilitators or creators of restrictions of certain behaviours (Páez & Scott, 2007; Hackney & Marchal, 2011; Kim, Rasouli, & Timmermans, 2017), social influence (Han, Arentze, Timmermans, Janssens, & Wets, 2011) and source of social capital (Carrasco et al., 2008a; Deutsch & Goulias, 2013). Aggregate measures describing social welfare and their interaction patterns – density, homophilia and heterogeneity – can be used as useful variables to describe travel behaviour (Carrasco & Miller, 2006; Sadri et al., 2015). Since a social contact is a measure of social well-being, it can therefore influence behaviour (Harvey & Taylor, 2000).

Specific characteristics of social networks have been found to influence travel behaviour. Closeness between members of a social network increases the frequency of face-to-face interactions (Chávez, Carrasco, & Tudela, 2017; Lin, 2016). Individuals with a bigger social network need to undertake more joint activities to maintain social relations (Lin & Wang, 2014). Social network size influences the number of trips and travel distances undertaken by individuals (van den Berg et al., 2013). On the other hand, the frequency of personal meetings decreases with physical distance between members of the same network (Lin, 2016).

Social networks are dynamic, and together with activity patterns, they evolve over time. They change as new connections are being created and old connection are getting lost, which directly affects activities scheduling (Arentze & Timmermans, 2008). Some of these dynamics are related to life-cycle events and are a relevant component to understand activity scheduling (Sharmeen, Arentze, & Timmermans, 2014).

The influence of the urban environment on social networks

The urban environment is the setting for many processes that shape social identity. In an increasingly urbanized world, the relationship between social interactions and the urban environment is of core importance and affects quality of life, health and well-being (EEA, 2009; WHO, 2009). Cities are expanding fast, which results in overloaded infrastructures, disorganization and social stress among its citizens (Butterworth, 2000). But space and social organization are related (Carmona, Heath, & Tiesdell, 2003) since societies exist in space and spatial characteristics influence social interactions. This influence can act both in direct and indirect ways. Directly it can facilitate interaction and communication among different socioeconomic groups and contribute to improving community cohesion, increasing the amount of physical and outdoor activity, while indirectly it can influence population health levels (Butterworth, 2000). Social networks are critical to the mental and emotional well-being of individuals, and their locality is particularly important (Halpern, 1995). The availability of services in a neighbourhood has a direct influence on the density of social networks and its perceptions. Public spaces can facilitate physical activity, promoting social interaction and cohesion and the build-up of social networks. People benefit from the vicinity of such places and access to them (Croucher, Myers, & Bretherton, 2008; De Vries, Verheeij, & Groenewegen, 2003). Public space provides an opportunity for shared use, interaction and activity in heterogeneous societies, contributing to the cohesion of communities. Vibrant communities contribute to increasing levels of social interaction, encouraging further the growth of social capital.

People are drawn to and spend more time in public spaces that are interesting and comfortable. The lack of public space, spatial segregation and quality of the built environment influence social interaction and the use of space. The quality and the safety of the built environment and its features are of crucial importance for the social networks and the social life, providing interaction opportunities and strengthening social ties.

Social networks begin in the neighbourhood. Jacobs (1961) observed that in neighbourhoods that maximize informal contacts among residents, street crime is reduced and people express greater happiness with their surroundings. The design of the neighbourhood, the proximity to public space and mix-use resources in them are of critical importance, and their design is a powerful tool to improve the social interactions among its inhabitants. Since the design and planning of urban spaces have social objectives, physical planning always has social consequences (Halpern, 1995). This means that planning decisions at various scales and levels have strong implications on the social environment.

Cities contribute to the generation of social interactions, which may occur in several ways and take place in different locations and settings (Bettencourt, 2013; Krafta, 2013). Transportation is closely associated with the everyday activities occurring in a community. The need for transportation derives from the need of people to be at locations where specific events occur, such as work, education, social businesses and so on. People travel to see other people, visit family or attend social/cultural/sports events. All cases require physical presence, which is considered to be the core of social interactions (Urry, 2002).

Social networks and social environment have strong spatial implications and are related to the built environment. Social processes are embedded in space and have strong implications for individual behaviour (McDonald, 2007). Cities are places that facilitate interactions between individuals with similar interests (Calthorpe & Fulton, 2001). The built environment determines the social environment and influences social contacts (Wang & Lin, 2013).

Social networks, defined as "a set of people or groups of people with some pattern of interactions or ties between them", represent individuals and their relationships (Newman, Watts, & Strogatz, 2002, p. 2566). Through social interactions, people exchange information about activity-travel choice options, based often on personal needs, preferences and attributes, which, in turn, formulate new or adjust existing social links. The development of these links and the maintenance of social networks involve travel and require the appropriate spatial support in the form of transportation services and infrastructure (Arentze & Timmermans, 2008).

The quality of life within cities is dependent on the human interaction with the urban environment (Das, 2008). The satisfaction level of public spaces, which are part of the urban environment, is directly affected by what is seen and how it is viewed (Lubis & Primasari, 2012). Communicating to others, through different media, individual experiences, opinions and suggestions about the urban environment, generates feedback. This feedback can be considered as part of social communication, which is among the most significant concepts taken up in connection with the urban environment, and refers to the transfer of information, thoughts and human behaviours between individuals (Koleini Mamaghani, Parvandar Asadollahi, & Mortezaei, 2015). Feedback allows complex systems to become adaptive (Johnson, 2001), meaning that cities develop and operate as adaptive self-organized systems, based on citizens' interactions with neighbours and the acknowledgement of patterns.

Understanding social networks, as part of spatial networks, can help the prediction of social activities schedules, the identification and explanation of travel patterns and consequently, the forecast of the demand for urban facilities and configurations (Ronald, Arentze, & Timmermans, 2009). This means that, apart from understanding with whom an activity will be performed, it is also very important to study "where" this activity will be carried out, incorporating in this case, social networks into activity-travel behaviour (Carrasco & Miller, 2006). The urban environment facilitates or may constrain the features of the interactions among people of one or more social networks. For groups of people with low incomes, the built environment can offer a space that enhances communication, networking and interchange (Lima, Carrasco, & Rojas, 2013). At the same time, cities accommodate different cultures in relatively restricted spaces and involve people belonging to multiple social networks (Knox & Pinch, 2010).

Different features of the urban environment have different effects on social networks and social interaction. Mixing public facilities and other uses (e.g. residential) facilitates face-to-face social interactions and the use of public space (van den Berg et al., 2014). Sprawl, decentralization, fragmentation and the resulting longer commutes reduce the potential for social interaction and face-to-face contacts (Farber & Li, 2013). The reduction of transport costs has implied that social networks will have bigger geographies and at the same time allow people to be more selective in their contacts, since they will not necessarily engage with their neighbours (Axhausen, 2007). Also, the increasing importance of leisure and social travel implies that locations will reflect the joint preferences of more than one person (Axhausen, 2007), implying a stronger effect from the urban environment characteristics.

The urban environment, which is considered to be a rather multidimensional concept, can be defined as a combination of urban design, land use and the transportation system, encompassing also human activities in the physical environment (Ngo, 2015). According to Handy, Boarnet, Ewing and Killingsworth (2002), five dimensions of the urban environment are recognized:

1   Density and intensity of development, measuring the amount of activity in an area, usually being defined as population, employment or buildings and measured by the number of people and/or jobs in a certain area
2   Mix of land uses, which refers to the different land uses within a given area. This dimension will include other land uses and urban activities like commerce, public facilities and services, parks, offices and other land uses.

3   Connectivity of the street network, regarding the directness or availability of alternative routes from one point to another within the street network

4   Scale of streets, determined by the space along a street as bounded by buildings or trees, walls and so forth

5   Aesthetic qualities of a place, contribute to its attractiveness or appeal. The design of buildings, ornamentation, landscaping and the availability of public amenities are among the factors that formulate the aesthetics of a place.

There is evidence that land use and transportation are interdependent, and different urban forms and land-use standards affect transport systems, addressing different travel patterns (Handy, Cao, & Mokhtarian, 2005). In parallel, transport supply affects firms and individuals location choices and, in the long term, influences also the structure of settlements (Lundqvist, 2003).

Urban form, addressed by the distances between locations where several activities take place and the provision of alternative transportation modes, creates the foundation, which facilitates some kinds of travel behaviour while discouraging others. It can be assumed that a higher population density signifies shorter average distances between residences, workplaces and service spaces, compared to a city that has scattered development patterns. It is also expected that in cities with a high population density, the quality of services of the public transport system is adequate (frequent departures, shorter walking access/egress distances), the streets are narrower and the availability of parking spaces is low (Naess, 2012). The findings of research carried out in Scandinavia, revealed that inhabitants of modern cities seem to give emphasis to the potentiality of choosing among facilities instead of proximity, demonstrating that the quantity of travel is affected by the location of the residence related to the concentration of facilities, rather than the distance to the closest single facility (Naess, 2012).

The vast number of different urban activities and inhabitants transforms cities into complex systems, which have to provide adequate and appropriate accessibility to urban spaces (Batty, 2012). The morphology of the street network is formed by the embedded spatial pattern of a city, that is land use and street topology, and linked also to dynamic attributes, such as congestion, accessibility and travel demand. Street networks are the most significant enablers of movements in the urban environment, allowing residents to navigate different functional features of a city and capturing the complex interactions between people (Lee, Barbosa, Youn, Holme, & Ghoshal, 2017). A healthy city is built on strong networks, and urban planning affects people's ability to socialize with others (Gilchrist, 2009; Farber & Li, 2013). Still, less socialization may also lead to a stagnant mobility, however not necessarily because of the lack of accessibility but due to a decreased requirement for third places to socialize (Rosenbaum, 2006).

A third place may be defined as a place other than the main spatial anchors of individuals, home (first place) and work/school (second place); it is a public place where interaction with other people happens (Oldenburg, 1989). There is ample demonstration in the literature about the significant role that third places have in the social life of specific groups of inhabitants, such as the elderly. A study conducted by Rosenbaum (2006), showed that third places cover the need of the elderly for companionship and emotional support, in addition to everyday consuming needs. These findings emphasize that modern cities should establish the appropriate physical environment, which can provide opportunities to all citizens for social interaction.

Public open spaces and in particular green spaces (which could be classified as third places) are also features of the physical environment that allow people to meet and interact with others within the context of the whole community, including family relationships, local social connections, cultural groupings and groups with common interests/habits (Holland, Clark, Katz, & Peace, 2017). The better a city facilitates such spaces, the more it contributes to the cohesion of communities and consequently to the adoption of the concept "architecture of community", which, in turn, improves citizens' quality of life (Holland et al., 2017; Khansari, Mostashari, & Mansouri, 2013), and supports collective life. The latter is considered to be a good opportunity to get away from the daily problems and stresses and to widen social interactions by meeting people from different groups, increasing sociability and creating an active and lively atmosphere (Marcus & Francis, 1998).

It is evident that urban green and urban ecosystem services are beneficial to citizens, and such a service is, also, urban cultivation, which promotes social and environmental interactions and sets the foundation of a city to provide well-being and place attachment for all citizens, independent their age, gender, culture, ethnicity socioeconomic and education status (Häkkinen et al., 2012). This means that a modern city with attractive shared spaces can reach social sustainability, including social interactions additionally to equity, solidarity and quality-of-life issues.

But is it not just the physical characteristics or urban environment that affect social networks and contacts; functional characteristics also have an influence. Since the study of Appleyard & Lintell (1972), it has been known that traffic conditions affect the intensity of social contacts and therefore the characteristics of local social networks. Major roads have divisive effects on communities resulting in fewer journeys on foot, affecting particularly older people and children (Mindell & Karlsen, 2012). Higher traffic volumes inhibit the use of streets as places of socialization and are related to the inhibition to use streets for social purposes (Mindell & Karlsen, 2012). The easiness to use and access shared spaces influences the sense of community (Kearney, 2006). Advocates of the New Urbanism have argued that it contributes to an increased sense of community and reduces the perception of isolation and increases social interaction between residents (Kim, 2007). Some research in the 1980s questioned these claims, but more recent works have found that residents in areas planned according to the principles of New Urbanism were found to have more social contacts and to engage in more outdoor activities (Brown & Cropper, 2001). Also, residents in more urban areas have a higher propensity to engage in more frequent public space social interactions (van den Berg et al, 2014). Environments that are socially tightly knit afford the opportunities to meet others without prior planning (Axhausen, 2007). But they also increase the social capital of individuals (Larsen, Axhausen, & Urry, 2006). The shift from spatially dense and socially overlapping networks reduces the chances of overlapping between different social networks, since their links are dispersed into space (Larsen et al., 2006).

People living in deprived conditions and in deprived areas tend to have their social networks restricted to people in similar situations; this contributes to the difficulties of these people to improve both their economic and social conditions (Calthorpe & Fulton, 2001). Evidence was found of a relationship between social network quality and health conditions (Mindell & Karlsen, 2012).

With the emergence of ICT and social media, several people have assumed that physical interactions could be replaced by technology, leading to a complete disconnection of the spatial and social dimensions (Calthorpe & Fulton, 2001; Humphreys, 2010). Social media and the use of mobile devices could facilitate social interaction in public spaces, but they could also increase parochialism by increasing the sense of commonality among groups of people (Humphreys, 2010). It can lead to homophilious tendencies rather than extending and bridging social circles (Humphreys, 2010) and to social selectivity (Axhausen, 2007). Social diversity of Internet users in public spaces is lower than that of regular users of these spaces, but it contributes to a broader participation in the public sphere (Hampton, Livio, & Sessions Goulet, 2010). Location-based services and mobile social networks affect the way individuals perceive urban spaces, but they also connect to other people in those places (de Souza e Silva & Frith, 2010).

Social networks and different appropriations of the urban space

When people have the opportunity to share their intentions, beliefs and behaviours, they increase the probability of repetition by a network of other users with similar intentions, beliefs and behaviours. Therefore, a single attitude shared through a social network can influence the activities of a broader group of individuals. Sharing of such information, either from face-to-face or digital interactions, creates social influence.

Conformity behaviour is the type of social influence that consists of mimicking the behaviour of others (Kim et al., 2017). People usually change their decisions to mimic the behaviour of other individuals in order to match attitudes, beliefs and activities. This phenomenon follows the social norms of social networks. These norms are usually linked with the efficient achievement of certain goals, the acceptance of the network members and/ or the maintenance of a positive self-concept (Cialdini & Goldstein, 2004). This is an endogenous effect implying that the propensity to behave in a certain manner varies depending on the behaviour of the reference group (Manski, 1993).

Still, copying each other has been the driving force behind the spread of activities in the urban fabric. Although the research between social and spatial interactions is relatively scarce (Carrasco & Miller, 2007), there is an emerging need to address all relevant social linkages and, at the same time, address the impact social networks may have in the urban environment and the activities taking place there.

Certain activities take place at certain locations based on urban environment characteristics and transportation performance and accessibility. Still, apart from spatial planning, social networks may influence the actual use of the urban space, without necessarily changing the built environment.

The use of public spaces (i.e. parks) for political speeches, the elderly gathering at playgrounds, the use of athletic stadiums for concerts, benches at isolated urban areas used as beds for the homeless, abandoned warehouse buildings hosting illegal activities, the use of public roads for demonstrations and so forth are indicative cases of social networks influencing the activities taking place in the urban fabric, while the built environment remains the same.

Examining the influence of social networks on selecting activity location seems to be substantial, considering that limited social interaction may lead to social exclusion and urban isolation. Schönfelder & Axhausen (2002) linked social exclusion with urban exclusion. Cass, Shove, and Urry (2005) linked the space dynamics with the network size, stating that restricted activity space offers limited social interactions and, therefore, the network cannot easily be extended (Cass et al., 2005).

Early studies considered that an individual's daily path through time and space shapes and is shaped by the person's social network, describing the social network as a time-space map of repeated social interactions (Fischer, Jackson, Stueve, Gerson, & Jones, 1977; Willmott, 1986; Rowe & Wolch, 1990).

Social networks are also linked with social activity-travel generation. Significant research has been made on the social influence of travel decisions (Axhausen, 2008; Schlich, Schönfelder, Hanson, & Axhausen, 2004), proving that the nature of the travel behaviour is highly influenced by properties of personal social networks. Kim et al. (2017) argue that social interaction has been more thoroughly studied than social influence; the latter has focused mainly on studying conformity behaviour. Information exchange and attitude formation may also influence the acceptance and adaptation of new technology and mobility concepts (e.g. electric cars, car and ride-sharing services and mobility-as-a-service concerns; Xiao & Lo, 2016).

In the era of ICT, the recent research interest has been attracted by digital social networks (DSN). DSNs are part of our daily life, proven to have great influence in numerous daily activities (consuming, travelling, politics etc.), meeting people's daily needs and changing their temporal behaviour and activities (Gao, Tang, & Liu, 2014; Liang, Zhao, & Xu, 2014). Regarding their influence on spatially located activities, the location-based service (LBS; users' check-in in particular locations) have expanded the research of geographic information recognition and activity mapping, playing an increasingly important role on urban space activity research (Rösler & Liebig, 2013).

Social networks act by transforming each individual or group of individuals into multiplying agents of information and knowledge about urban spaces, influencing their perception, occupation and use.

Today, Foursquare travellers are able to interact directly with locals once arriving at a specific place; Facebook and Twitter check-ins inform, in real time, the other members of a network about activities taking place in specific geolocations; Trip Advisor users (i.e. tourists) may increase the attractiveness of specific places based on reviews of other members. These social networks may influence to a great extent the activities taking place in the urban environment, creating new dynamics on interactions among different social networks and creating new norms and trends for urban life.

Pereira, Florentino and Rocha (2013), argued that the DSNs play a cooperative and aggregator role, contributing to developing strategies to establishing contact with the urban spaces and equipment that are often unknown or ignored. Through the DSN it is also possible to diversify the use of streets and means of transportation, modifying traffic flows through the space of the city. DSNs introduced another set of interactions between the citizen and the urban space. What is worth mentioning is how the DSNs influence mediates the dynamics in choice set composition and spatial awareness. DSNs have the support of additional means of land and activities representation such as images, sounds and videos that are able to improve the perception of the urban environment. The geolocation of a citizen (localized now) contextualizes searches, introducing new meaning to the physical space.

Conclusion

This chapter reviewed the work focusing on the interrelationships between travel behaviour, social networks and the urban environment. Research has shown that social networks influence travel behaviour by being (1) sources of information spill-overs, (2) creators of restrictions, (3) medium of social influence and (4) sources of social capital. However, since social networks and the social environment have relevant spatial implications, it could be argued that the urban environment, by influencing specific characteristics of social networks, also influences travel behaviour in ways other than changing the relative spatial distribution of travel opportunities and the costs of using different travel modes. The built environment has both direct and indirect influences on social interactions. The built environment can facilitate interaction and communication among different groups and contribute to improving community cohesion and increase the amount of outdoor activity. Different features of the urban environment have different effects on social networks and social interaction. Sprawl and decentralization reduce the potential for social interaction, and the reduction of transport costs helped the creation of social networks with bigger geographies. But it is not just the physical characteristics or urban environment that affect social networks and contacts; functional characteristics also have an influence. With the emergence of ICT and social media, several people have assumed that physical interactions could be replaced by technology, leading to a complete disconnection of the spatial and social dimensions. Although social media and ICTs could facilitate social interaction, they could also trigger parochialism and social selectivity. This last aspect has societal wider implications that go beyond travel behaviour and reinforce the importance of the urban environment in sustaining and supporting a healthy and diverse social environment.

References

Appleyard, D., & Lintell, M. (1972). The environmental quality of city streets: The residents' viewpoint. *Journal of the American Planning Association, 38*(2), 84–101. https://doi.org/10.1080/01944367208977410

Arentze, T., & Timmermans, H. (2008). Social networks, social interactions, and activity-travel behavior: A framework for microsimulation. *Environment and Planning B: Planning and Design, 35*(6), 1012–1027. https://doi.org/10.1068/b3319t

Axhausen, K. W. (2007). Activity spaces, biographies, social networks and their welfare gains and externalities: Some hypotheses and empirical results. *Mobilities, 2*(1), 15–36. https://doi.org/10.1080/17450100601106203

Axhausen, K. W. (2008). Social networks, mobility biographies, and travel: Survey challenges. *Environment and Planning B: Planning and Design, 35*, 981–996. doi:10.1068/b3316t

Batty, M. (2012). Building a science of cities. *Cities, 29*, S9–S16. https://doi.org/10.1016/j.cities.2011.11.008

Bettencourt, L. (2013). The origin of scaling in cities. *Science, 340*, 1438.

Brown, B. B., & Cropper, V. L. (2001). New urban and standard suburban subdivisions: Evaluating psychological and social goals. *Journal of the American Planning Association, 67*(4), 402–419. https://doi.org/10.1080/01944360108976249

Butterworth, I. (2000). *The relationship between the built environment and well-being: Opportunities for health promotion in urban planning.* Melbourne: Victorian Health Promotion Foundation.

Calthorpe, P., & Fulton, W. (2001). *The regional city: Planning for the end of sprawl.* Washington, DC: Island Press.

Carmona, M., Heath, Oc, T., & Tiesdell, S. (2003). *Public places – urban spaces: The dimensions of urban design.* Oxford: Architectural Press.

Carrasco, J. A., Hogan, B., Wellman, B., & Miller, E. J. (2008a). Agency in social activity interactions: The role of social networks in time and space. *Tijdschrift Voor Economische En Sociale Geografie (Journal of Economic & Social Geography), 99*(5), 562–583. https://doi.org/10.1111/j.1467-9663.2008.00492.x

Carrasco, J. A., Hogan, B., Wellman, B., & Miller, E. J. (2008b). Collecting social network data to study social activity-travel behavior: An egocentric approach. *Environment and Planning B: Planning and Design, 35*(6), 961–980. https://doi.org/10.1068/b3317

Carrasco, J. A., & Miller, E. J. (2006). Exploring the propensity to perform social activities: A social network approach. *Transportation, 33*(5), 463–480. https://doi.org/10.1007/s11116-006-8074-z

Carrasco, J. A., & Miller, E. J. (2007). The social dimension in action: A multilevel personal networks model of social activity frequency. *Proceedings of the TRB 2007 Annual Meeting.*

Cass, N., Shove, E., & Urry, J. (2005). Social exclusion, mobility and access. *The Sociological Review*, 539–555.

Chávez, Ó., Carrasco, J. A., & Tudela, A. (2017). Social activity-travel dynamics with core contacts: Evidence from a two-wave personal network data. *Transportation Letters*, 1–10.

Cialdini, R. B., & Goldstein, N. J. (2004). Social influence: Compliance and conformity. *Annual Review of Psychology, 55,* 591–621. doi:10.1146/annurev. psych.55.090902.142015

Croucher, K., Myers, L., & Bretherton, J. (2008). *Greenspace Scotland Research Report: The links between greenspace and health: A critical literature review.* Stirling, UK: Greenspace Scotland.

Das, D. (2008). Urban quality of life: A case study of Guwahati, Springer Science and Business Media B.V. *Social Indicators Research, 88,* 297–310.

de Souza e Silva, A., & Frith, J. (2010). Locative mobile social networks: Mapping communication and location in urban spaces. *Mobilities, 5*(4), 485–505. https://doi.org/10.1 080/17450101.2010.510332

Deutsch, K., & Goulias, K. G. (2013). Decision makers and socializers, social networks and the role of individuals as participants. *Transportation, 40*(4), 755–771. https://doi. org/10.1007/s11116-013-9465-6

De Vries, S., Verheeij, R., & Groenewegen, P. (2003). Natural environments? An exploratory analysis of the relationship between greenspace and health. *Environment and Planning A, 35,* 1717–1731.

EEA. (2009). Ensuring quality of life in European cities and towns. *EEA Report no 5.* Office for Official Publications of the EU, Luxembourg.

Farber, S., & Li, X. (2013). Urban sprawl and social interaction potential: An empirical analysis of large metropolitan regions in the United States. *Journal of Transport Geography, 31,* 267–277. https://doi.org/10.1016/j.jtrangeo.2013.03.002

Fischer, C. S., Jackson, R. M., Stueve, C. A., Gerson, K., & Jones, L. M. (1977). *Networks and places: Social relations in the urban setting.* New York, NY: The Free Press.

Gao, H., Tang, J., & Liu, H. (2014). Personalized location recommendation on location-based social networks. In *Proceedings of the 8th ACM Conference on Recommender Systems* (pp. 399–400). Silicon Valley: ACM, 6–10 October 2014.

Gilchrist, A. (2009). *The well-connected community: A networking approach to community development* (2nd ed.). London: Policy Press.

Hackney, J., & Marchal, F. (2011). A coupled multi-agent microsimulation of social interactions and transportation behavior. *Transportation Research Part A: Policy and Practice, 45*(4), 296–309.

Hägerstrand, T. (1970). What about people in regional science? *Papers in Regional Science, 24*(1), 7–24. https://doi.org/10.1111/j.1435-5597.1970.tb01464.x

Häkkinen, A., Kaasalainen, H., Laitinen, V., Nollet, C., Sagizbaeva, O., Sannemann, C., & Vuorinen, S. (2012). Attractive shared spaces in social sustainability and quality of life. *The multidisciplinary report.* Environmental special course at the Helsinki University Centre for Environment: Urban Ecosystem Services.

Halpern, D. (1995). *Mental health and the built environment.* London: Taylor & Francis.

Hampton, K. N., Livio, O., & Sessions Goulet, L. (2010). The social life of wireless urban spaces: Internet use, social networks, and the public realm. *Journal of Communication, 60*(4), 701–722. https://doi.org/10.1111/j.1460-2466.2010.01510.x

Han, Q., Arentze, T., Timmermans, H., Janssens, D., & Wets, G. (2011). The effects of social networks on choice set dynamics: Results of numerical simulations using an agent-based approach. *Transportation Research Part A: Policy and Practice, 45*(4), 310–322. https://doi.org/10.1016/j.tra.2011.01.008

Handy, S. L., Boarnet, M. G., Ewing, R., & Killingsworth, R. E. (2002). How the built environment affects physical activity: Views from urban planning. *American Journal of Preventive Medicine, 23*(2), 64–73. https://doi.org/10.1016/S0749-3797(02)00475-0

Handy, S. L., Cao, X., & Mokhtarian, P. (2005). Correlation or causality between the built environment and travel behavior? Evidence from Northern California. *Transportation Research Part D: Transport and Environment, 10*(6), 427–444.

Harvey, A. S., & Taylor, M. E. (2000). Activity settings and travel behaviour: A social contact perspective. *Transportation, 27*(1), 53–73. https://doi.org/10.1023/a:1005207 320044

Holland, C., Clark, A., Katz, J., & Peace, S. (2017). *Social interactions in urban public spaces.* York: Joseph Rowntree Foundation.

Humphreys, L. (2010). Mobile social networks and urban public space. *New Media & Society, 12*(5), 763–778. https://doi.org/10.1177/1461444809349578

Jacobs, J. (1961). *The death and life of great American cities.* New York, NY: Random House.

Johnson, S. (2001). *Emergence: The connected lives of ants, brains, cities, and software.* New York, NY: Scribner.

Kearney, A. R. (2006). Residential development patterns and neighborhood satisfaction: Impacts of density and nearby nature. *Environment and Behavior, 38*(1), 112–139. https://doi.org/10.1177/0013916505277607

Khansari, N., Mostashari, A., & Mansouri, M. (2013). Impacting sustainable behaviour and planning in smart city. *International Journal of Sustainable Land Use and Urban Planning, 1*(2), 46–61. ISSN 1927–8845

Kim, J. (2007). Perceiving and valuing sense of community in a New Urbanist development: A case study of Kentlands. *Journal of Urban Design, 12*(2), 203–230. https://doi.org/10.1080/13574800701306286

Kim, J., Rasouli, S., & Timmermans, H. J. P. (2017). Social networks, social influence and activity-travel behaviour: A review of models and empirical evidence. *Transport Reviews,* 1–25. http://doi.org/10.1080/01441647.2017.1351500

Knox, P., & Pinch, S. (2010). *Urban Social Geography: An Introduction* (6th ed.). London: Pearson Education Limited. ISBN: 978-0-273-71763-8.

Koleini Mamaghani, N., Parvandar Asadollahi, A., & Mortezaei, S.-R. (2015). Designing for improving social relationship with interaction design approach. *Procedia – Social and Behavioral Sciences, 201*(2015), 377–385.

Krafta, R. (2013). On scaling functionality in urban form. ArXiv Physics and Society. https://arxiv.org/abs/1311.2462

Larsen, J., Axhausen, K. W., & Urry, J. (2006). Geographies of social networks: Meetings, travel and communications. *Mobilities, 1*(2), 261–283. https://doi.org/10.1080/17450100600726654

Lee, M., Barbosa, H., Youn, H., Holme, P., & Ghoshal, G. (2017). Morphology of travel routes and the organization of cities. *Nature Communications, 8*, 2229.

Liang, X., Zhao, J. C., & Xu, K. (2014). Analysis of social ties and check-ins in location-based social networks. *Science and Technology Review, 32*(11), 43–48.

Lima, C., Carrasco, J. A., & Rojas, C. (2013, July). Built environment and travel for social interactions: case studies in different urban contexts. 13th WCTR, Rio de Janeiro, Brazil, July 15–18, 2013.

Lin, T. (2016). *Personal social networks, neighborhood social environments and activity-travel behavior.* Hong Kong: Open Access Theses and Dissertations.

Lin, T., & Wang, D. (2014). Social networks and joint/solo activity – travel behavior. *Transportation Research Part A: Policy and Practice, 68.* kötet, 18–31.

Lubis, B. U., & Primasari, L. (2012). The relationship between people and urban screen in an urban space. *Procedia – Social and Behavioral Sciences, 42*(2012), 223–230.

Lundqvist, L. (2003). Land-use and travel behaviour: A survey of some analysis and policy perspective. *EJTIR, 3*(3), 299–313.

Manski, C. F. (1993). Identification of endogenous social effects: The reflection problem. *The Review of Economic Studies, 60*(3), 531–542. doi:10.2307/2298123

Marcus, C. C., & Francis, C. (1998). *People places: Design guidelines for urban open spaces.* New York, NY: Van.

McDonald, C. N. (2007). Travel and the social environment: Evidence from Alameda County, California. *Transportation Research Part D: Transport and Environment, 12*(1), 53–63. https://doi.org/10.1016/j.trd.2006.11.002

Mindell, J. S., & Karlsen, S. (2012, April). Community severance and health: What do we actually know? *Journal of Urban Health.* https://doi.org/10.1007/s11524-011-9637-7

Naess, P. (2012). Urban form and travel behavior: Experience from a Nordic context. *The Journal of Transport and Land Use, 5*(2), 21–45.

Newman, M. E. J., Watts, D., & Strogatz, S. H. (2002). Random graph models of social networks. *Proceedings of the National Academy of Sciences, 99*, 2566–2572.

Ngo, D. (2015). *Influencing travel behavior in urban areas: A study on the factors of reaching sustainable mobility through travel behavior.* Master thesis, Aalborg University.

Oldenburg, R. (1989). *The great good place: Cafes, coffee shops, community centers, beauty parlors, general stores, bars, handouts, and how they get you through the day.* New York, NY: Paragon House.

Páez, A., & Scott, D. M. (2007). Social influence on travel behavior: A simulation example of the decision to telecommute. *Environment and Planning A, 39*(3), 647–665. https://doi.org/10.1068/a37424

Pas, E. I. (1985). State of the art and research opportunities in travel demand: Another perspective. *Transportation Research Part A: General, 19*(5–6), 460–464. https://doi.org/10.1016/0191-2607(85)90048-2

Pereira, G. C., Florentino, P. V., & Rocha, M. C. F. (2013). City as a social network- Brazilian examples. In C. Ellul, S. Zlatanova, M. Rumor, & R. Laurini (Eds.), *Urban and regional data managements: UDMS Annual 2013.* London: Taylor & Francis Group. ISBN 978-1-138-0063-6

Ronald, N. A., Arentze, T. A., & Timmermans, H. J. P. (2009, July). An agent-based framework for modelling social influence on travel behavior. 18th World IMACS/MODSIM Congress, July 13–17, 2009, Cairns, Australia.

Rosenbaum, M. S. (2006). Exploring the social supportive role of third places in consumers' lives. *Journal of Service Research, 9*(1), 59–72.

Rösler, R., & Liebig, T. (2013). Using data from location based social networks for urban activity clustering. In *Geographic information science at the heart of Europe* (pp. 55–72). New York, NY: Springer International Publishing.

Rowe, S., & Wolch, J. (1990). Social networks in time and space: Homeless women in skid row, Los Angeles. *Annals of the Association of American Geographers, 80*, 184–204.

Sadri, A. M., Lee, S., & Ukkusuri, S. V. (2015). Modeling social network influence on joint trip frequency for regular activity travel decisions. *Transportation Research Record: Journal of the Transportation Research Board, 2495*, 83–93. https://doi.org/10.3141/2495-09

Salomon, I. (1985). Telecommunications and travel: Substitution or modified mobility? *Journal of Transport Economics and Policy, 19*(3), 219–235. https://doi.org/10.1080/09595238100185051

Schlich, R., Schönfelder, S., Hanson, S., & Axhausen, K. W. (2004). Structures of leisure travel: Temporal and spatial variability. *Transport Reviews, 24*(2), 219–237.
doi:10.1080/0144164032000138742

Schönfelder, S., & Axhausen, K. W. (2002). Measuring the size and structure of human
activity spaces ñ the longitudinal perspective. *Arbeitsbericht Verkehrs- und Raumplanung, 135*, Institut f,r Verkehrsplanung und Transportsysteme (IVT), ETH Z,rich, Z,rich.

Sharmeen, F., Arentze, T., & Timmermans, H. (2014). An analysis of the dynamics of
activity and travel needs in response to social network evolution and life-cycle events:
A structural equation model. *Transportation Research Part A: Policy and Practice, 59*.
kötet, 159–171.

Urry, J., (2002). Mobility and proximity. *Sociology – The Journal of the British Sociological Association, 36*, 255–274.

van den Berg, P., Arentze, T., & Timmermans, H. (2013). A path analysis of social networks, telecommunication and social activity-travel patterns. *Transportation Research
Part C: Emerging Technologies, 26*, 256–268. https://doi.org/10.1016/j.trc.2012.10.002

van den Berg, P., Kemperman, A., & Timmermans, H. (2014). Social interaction location
choice: A latent class modeling approach. *Annals of the Association of American Geographers, 104*(5), 959–972. https://doi.org/10.1080/00045608.2014.924726

Wang, D., & Lin, T. (2013). Built environments, social environments, and activity-travel
behavior: A case study of Hong Kong. *Journal of Transport Geography, 31*, 286–295.
https://doi.org/10.1016/j.jtrangeo.2013.04.012

WHO. (2009). *Zagreb Declaration for healthy cities: Health and equity in all local policies*. Copenhagen: WHO regional Office for Europe.

Willmott, P. (1986). *Social networks, informal care and public policy*. London: Policy
Studies Institute.

Xiao, Y., & Lo, H. K. (2016). Day-to-day departure time modeling under social network
influence. *Transportation Research Part B: Methodological, 92*, 54–72. doi:10.1016/j.
trb.2016.05.006

10 Modelling the city in dialogue with new social media and modes of travel behaviour

Antònia Casellas and Itzhak Omer

Introduction

Determining how the use of social media and digital social networks (DSNs; e.g. Facebook, Twitter, Gowalla, and Foursquare data) influences the definition of a *city* is a necessary inquiry, from both academic and policy perspectives. It is also an elusive topic that presents a dilemma resulting from the interpretation of a new social phenomenon in constant transformation and the fact that, to date, scholars and practitioners have not been able to agree on a unique and valid definition of what attributes define a city.

The concepts of *city* and *urban* are closely related. Nevertheless, as Fishman (1987) points out, depending on the perspective of scholars and practitioners, city attributes and features can vary significantly. Ebenezer Howard (1898), Frank Lloyd Wright (1932), and Le Corbusier (1933) all reacted to the same phenomenon: the grimy features of industrial cities; nonetheless, they designed very different ideal urban environments, ranging from a small city surrounded by nature to a densely populated and efficiently managed city. Moreover, together with the existence of different designs and plans, from past to present, cities have remained centres of political, economic and cultural power (Hall, 1988). But most of all, cities provide the conditions for the formation of social interaction among people. Accordingly, many efforts have been devoted to the question of how political, economic, cultural, and technological factors, as well as urban design aspects, are related to social interaction in the city.

In the era of globalization, some of the powerful images of what constitutes a city recall Jean Gottmann's definition of megalopolis (1961) as a colloidal mixture of rural and suburban landscapes; Saskia Sassen (2001) and Taylor's (2004) global and world cities as centres of command and control nodes of the global economy; and Florida, Gulden, and Mellonder (2007) mega-regions as clustered cities that operate as economic organizing units producing the bulk of the world's wealth and attracting a large share of creative people. In past decades, the new values and cultural systems, together with the technological potentials that the ICT offers, have contributed to another radical transformation in how we live and communicate through time and space in an urban environment. A city is no longer the place where we just live or work but the place where we travel, move and visit

(Poli, 2009). From this approach, this transformation in the use of the city's space seems to make many of the interpretative schemes of what a city is, or should be, quite obsolete.

The processes of transformation have directly affected the use of urban space and call for an effort to understand cities, in dialogue with new social media and modes of travel behaviour. As Slavoj Žižek, Milbank, and Davis (2009) point out about philosophy, redefining what most of the academic and political class identifies as "the problem" is always necessary. By the same logic, in the context of urban studies and policy, the impact of a DSN and its interaction with new patterns of mobility in space reopens the debate of how to interpret and define the city and to design urban models.

To explore this topic further and contribute to the existing debate, in this chapter, we focus first on the challenges of finding a definition of *city* that allows comparative studies and guides cross-national urban policies. Here, we pay special attention to the recent efforts of the European Union. Then, the following section addresses how urban models can be designed and applied in dialogue with social networks and new mobility patterns. In this part, we present the development of cities and urban modelling approaches, and then we discuss the relevance and potential of a DSN for urban modelling. The chapter concludes by highlighting some of the potential and limitations of a DSN in urban modelling and policy.

Contested definitions and attributes

In urban sociology, cities have been defined as being in opposition to what is considered rural (Frey & Zimmer, 2001). This classic approach finds its roots in Tönnies' (1979/1887) traditional distinction between Gemeinschaft (community) and Gesellschaft (society), as two different social ties with its set of roles, values and beliefs that differentiate rural life from urban life. This differentiation has become quite irrelevant because the traditional rural world no longer exists in a differentiated way with respect to the urban one. New technologies, high mobility, the globalization of information and cultural practices prevent the differentiation of a city person from a town or village one. However, as Poli (2009) points out, this vision has been a persistent conceptual tool in urban sociology that must be questioned further to advance the debate.

On the other hand, from a policy perspective, cities are often defined as administrative boundaries, which are enriched with statistical data information, the inclusion of historical reasons, the description of governmental decisions and measures of population density. Apart from the dominant administrative criterion, other standards that traditionally have contributed to the definition of a city in urban geography studies are based on the use of one or a combination of several criteria. They include (1) morphological criteria, that is, from the delimitation of the urbanized space; (2) functional criteria, considering mobility patterns and communication networks; (3) economic criteria, centred on the analysis of the characteristics of structures and economic patterns; and (4) lifestyle criteria, often in opposition to rural lifestyles. All of these criteria, individually or grouped in different variations, have been used and are still used to reach an understanding of cities (Frey & Zimmer, 2001; Bettencourt, 2013). Since the second decade of the 21st century, these traditional debates have been enriched by the contribution of big spatial data analysts.

The inclusion of DSN data to urban definitions brings a new approach to the many already existing methodologies, which could advance urban studies and policies. Nevertheless, considering that the exact definition of a city "depends on who you ask" (Deuskar, 2015), in comparative studies, most often a practical solution has been to rely again on administrative delimitation and statistical data gathered in accordance with the criteria that a specific country establishes. This strategy generates several challenges for scholars, planners and policy-makers when they establish comparative analysis across countries.

The main challenge for comparative urban studies is that often the territorial units are not comparable and could lead to a biased interpretation. This is a highly problematic challenge, but not uncommon. The European Union's Nomenclature of Territorial Units for Statistics, referred to by the French acronym NUTS II, which is used as the basis for regional development policies in the European Union, is also extremely heterogeneous. Metropolitan areas, tiny islands, large rural regions and even whole countries are considered comparable units of analysis as different countries classify them as NUTS II. Despite its lack of coherence, and as it has been noticed, this problematic classification is used as the basis for the EU regional development policies, which has clear implications for identifying and solving regional disparities in the European Union (Casellas & Galley, 1999). Overcoming the heterogeneity of cities' administrative boundaries has also become a much-needed priority for urban studies and policy.

In the European Union, the complexity of the urban phenomenon and the extreme inconsistencies among country classifications pushed policy-makers in 2011 to harmonize a definition of cities and their surrounding areas. The work was carried out by the European Commission's Directorate-General for Regional and Urban Policy (DG REGIO), Eurostat, and the Organisation for Economic Co-operation and Development (OECD).

The OECD–EU approach was based on three steps (Dijkstra & Poelman, 2012). The first step was to identify a contiguous or highly interconnected, densely inhabited urban core. For that purpose, the OECD established that a city consists of one or more local administrative units (LAU), where most of the population lives in an urban centre of at least 500 inhabitants per km^2, with a total population of at least 50,000 inhabitants. The second step was to group urban cores into functional urban areas (FUA), if more than 15% of the population of any of the cores commutes to work to the other cores. Finally, the third step was to define the commuting shed or "hinterland" of the FUA, based on travel-to-work community flows from surrounding municipalities to the urban core, following the same criteria of a minimum of 15% of its resident, employed population of any of the cores commuting to the urban core.

To define the urban cores, this methodology used population size cut-offs (50,000 or 100,000 people, depending on the country) and population density cut-offs (1,000 or 1,500 people per km^2), and selected the areas from which more than 15% of workers commute to the core.

This new OECD–EU definition of FUA provides a methodology that helps to compare FUA of similar size across countries, because it has the advantage that extends across administrative boundaries. From that, it establishes a typology with four categories of FUA: (1) small urban areas, with a population between 50,000 and 200,000; (2) medium-sized urban areas, with a population between 200,001 and 500,000; (3) metropolitan areas, with a population between 500,001 and 1.5 million; and (4) larger metropolitan areas, with a population of more than 1.5 million.

Applied to 30 OECD countries, the typology allows identification of 1,179 FUA of different sizes. It further identifies 281 metropolitan areas, allowing comparison across a range of indicators in metropolitan areas. The methodology, however, has its own clear limitations. The definition relies on population density to identify urban cores and travel-to-work flows to identify the hinterlands, whose labour market is highly integrated with the cores. So, it does not take into account other criteria, such as economic or lifestyle criteria and daily activities. So, different activity types that are more coherent and representative of movement flows might enhance the actual definition of FUA.

By identifying a contiguous or highly interconnected, densely inhabited urban core, and linking it to mobility patterns, the OECD–EU definition takes into account two traditional criteria: morphological criteria, that is from the delimitation of the urbanized space, and a variable within the functional criteria, as it includes mobility patterns. Nevertheless, this methodology excludes other features that long and well-established traditions of sociologists, urbanists, planners and geographers have identified as key to the essence of urban life, as well as the information provided by massive volumes of data generated by a DSN.

An aspect that has been linked to urban identity, and which is still highly relevant to the analysis of urban life, is the concept of urban vitality (see Delclòs-Alió & Miralles-Guash, 2018). By urban vitality, we refer to the community and spiritual values associated with the urban experience that allow social interaction among diverse groups. This is one of the attributes that Jane Jacobs highlights in her seminal 1961 book *The Death and Life of Great American Cities*. Her argument is built against urban renewal programmes in the United States and their negative effects on neighbourhood communities. Confronting Robert Moses's master plans for New York City, she identifies the interaction among strangers in public space as a key attribute of healthy and safe cities. Within this frame, the bustling sidewalks are places that generate a pedestrian environment, which makes a city safe. In the same vein, Lewis Mumford (1937, 1938) and Louis Wirth (1938) highlight the important capacity of the city to provide its residents with an urban experience that enhances human potential. In *What Is a City?* Mumford (2000/1937) gives a sociological answer to this question. Guiding the priorities of the planners, he asserts that what defines a city are the opportunities it offers to different social groups. In this sense, the dimensions of a city and its population size are not important, but the social relationships they help to achieve are important.

When placing these definitions in the context of the possibilities of interaction and communication offered by a DSN, we could argue that we are evolving towards a new period. In the era dominated by postmodern values, past individual identities considered indivisible and unique, as defined by professions, religious beliefs and nationalities, among other factors, have been broadly questioned. We have moved to understanding ourselves as multiple subjectivities decoded through language games, syncretic cultural practices and cultural relativism (Butler, 1990). As a result, we could argue that, parallel with our new subjectivities and practices, the city as a place that contains our activities has been transformed. We live in a flexibility and plurality of times, for both labour and leisure, in which ICT technologies allow us to communicate across space and inform people in real time about the multiple possible interactions with the built environment.

To move forward in this debate, we could understand the city from a systemic perspective and see how the process of urbanization is linked to a multidimensional global change (Harvey, 1989, 1996, 2000), in which ICT technologies play a crucial role. This approach requires breaking radically with inherited assumptions regarding the spatiality of the urban and engaging with the reconceptualization of the urban issue. As Harvey (1996) points out, social processes produce and reproduce new spatio-temporalities.

In the same line of thought, Brenner and Schmid (2014, 2015) insist that cities must be analysed as dynamically evolving sites. As such, they need to be understood as arenas and outcomes of broader processes of socio-spatial and socio-ecological transformation that go far beyond their spatial structure. Brenner (2014) reminds us that the urbanized conditions resulting from the development, intensification and worldwide expansion of late capitalism produce a terrain that extends beyond the zones of urban agglomeration that have monopolized the attention of urban researchers.

All this points to the need to explore in detail how urban models could adapt to new social media and mobility patterns, which will help better understand the city from a much-needed systemic perspective.

Urban models in dialogue with social networks and new mobility patterns

Urban models are essential to understanding the structure and dynamics of cities and testing the impact of changes in land use and transportation on the urban environment. Because of the increased involvement of a DSN in contemporary urban society, we suggest they can be relevant to urban modelling in two ways. First, they can serve as an appropriate, unique provider of data that can tell us how and when individuals use urban spaces. Second, as they have an effect on individuals' spatial behaviour, they should be considered in urban modelling. In the following section, we present the development of cities and urban modelling approaches, and then we discuss the relevance and potential of a DSN for urban modelling.

City development and urban models

The classic urban models suggested by the urban ecology approach, such as the concentric zone (Burgess, 1925), the sector model (Hoyt, 1939) and more recent examples (Ehlers, 2011), represent the spatial patterns of residential and non-residential land use together with their relative position within the city. These models of the industrial monocentric city were based mainly on socio-economic, "ecological" mechanisms, such as segregation of social groups that resulted in distinct functional and social urban areas, and the computation between land use in the best locations relative to the city centre, as also discussed in the urban economics framework. Following the development of new transport and ICTs, as well as processes related to local economies of agglomeration, modern and postmodern cities have been increasingly evolving into being more polycentric than what was represented in the classic models (Anas, Arnott, & Small, 1998). Polycentric cities are made up of many activity centres (commerce and public services) at various geographical scales, located outside traditional central business districts and include the formation of "edge cities" (Batty, 2008), cities around cities. These changes in the city's spatial structure are reflected in the movement flows between the activity centres and social areas. Whereas in the traditional monocentric cities the dominant flows were from suburbs to the city centre, in polycentric cities people are channelled by a range of travel needs across centres at different levels, with cross-commuting throughout and between cities becoming relatively more dominant (Batty, 2008). Urban space thus becomes much more disorderly, heterogeneous, complex and dynamic.

The first dynamic urban models were developed to simulate the movement flows and spatial-temporal changes in land-use distribution. These models were mainly of the spatial interaction type (Wilson, 1970), which represents movement flows between large-scale urban areas and diffusion models that simulate space-time changes, such as the development of the residential patterns of minority groups (Hagerstrand, 1967; Morrill, 1965). Cellular Automata models arrived later, during the 1990s, for modelling the dynamics of land-use patterns (White, Engelen, & Uljee, 1997, 2015). These computerized dynamic models require urban space to be represented by means of cellular spaces, occupied by different land uses or social groups. These dynamic models, however, are used mostly with large-scale aggregate data, primarily administrative areas; hence, they cannot, at least explicitly, consider the spatial behaviour and dynamics of individuals (Benenson & Torrens, 2004). As a result, they do not capture micro-scale functional and social activities, where the relationship between land-use distribution, movement and street network come together.

This situation has nonetheless changed significantly since the early 2000s as a result of improvements in the construction of geographical databases and geographic information system technology, making it possible to obtain high-resolution, geo-reference data regarding land-use distributions, as well as the built environment and socio-demographic properties (Benenson & Omer, 2003). The availability of such data allows us to refer to land use and social variables at the resolution of individual buildings and streets.

Modelling the association between street-network structure, land-use distribution and movement flow at the street level has been conducted mainly within the configurational space syntax approach (Hillier, Penn, Hanson, Grajewski, & Xu, 1993). According to this network-based approach, a street network's spatial configuration creates the potential for movement flows (i.e. "natural movement") throughout the urban street network, according to the centrality of each street at different scales. Moreover, as suggested by the concept "movement economy process" (Hillier & Penn, 1996), the configuration of the street network, through its interaction with pedestrian and vehicular movement flows, is largely related to the distribution of land use. This leads, in turn, to the emergence of networks of linked centres containing retail land use and other movement-driven activities by means of economic agglomeration (Vaughan, Jones, Griffiths, & Haklay, 2010). Network-based urban models, which are based on the street network's centralities in addition to land use and socio-economic variables, have been used for exploring urban movement (Jiang, 2009; Omer, Rofe, & Lerman, 2015; Ozbil, Peponis, & Stone, 2011) and for predicting pedestrian and vehicular movement flows, especially in transport planning (e.g. Desyllas, Duxbury, Ward, & Smith, 2003; Lerman, Rofè, & Omer, 2014). And yet, network-based urban models refer to limited aspects of individuals' spatial behaviour, that is to the distance types people use to calculate the shortest routes to their destinations from within the network (metric, topological and angular distance) and to the relative attractiveness of land use, which tends to be defined according to objective criteria, such as the number and size of non-residential buildings.

The past 30 years have witnessed the emergence of urban complexity theories that refer to cities as complex self-organizing systems. According to these theories, cities originally emerged from the space-time interactions between the many urban agents (i.e. individuals, families, households, firms, and other entities) that act and interact in the city (Batty, 2013; Portugali, 1999). These activities and interactions give rise to the urban multilevel functional and spatial structure that, in turn, affects the agent's cognition, behaviour, movement and action in the city in the form of circular causality (Batty, 2013; Portugali, 2011). So, according to these theories, cities emerged not from any top-down action but from the bottom up. Such ideas encourage the modelling of urban dynamic systems in terms of individuals, not aggregate populations. And indeed, these complexity and self-organization theories, as well as changes in computational power and availability of geo-reference data, have inspired the development of agent-based (AB) urban models (Batty, 2005). Moreover, unlike aggregate-based modelling (i.e. aggregation by variables and/or area), AB models can represent the dynamics of individuals with different goal-oriented behaviours, cognitions of space and personal interactions (Jiang & Jia, 2011; Torrens et al., 2012). The same models are also able to contend with the simultaneous dependence of movement flow on the street-network structure and land use (Batty, 2005). In consequence, many AB models have been applied to simulating movement flows and land-use distributions (Ferguson, Fridrisch & Karimi, 2012; Omer & Kaplan, 2017). With the increasing availability of geo-referenced "big data", it is reasonable to assume that the application of AB modelling is expected to broaden (Goodchild, 2007). However, despite the dramatic change in the availability of *objective* functional and behavioural databases, *subjective* data on individuals' intentions, perceptions, and preferences with respect to different places in the city are much less available.

In the following subsection, we discuss the potential of DSN data to enhance urban modelling trends, particularly models aimed at exploring the self-organized urban process derived from the relationships maintained between an agent's spatial behaviour, movement-flow patterns and the spatial structure of functional and social activities in cities.

DSN data: the potential for urban modelling

DSN data are contributed by individual users and generally include social-media check-ins (e.g. in business and services), points of interest, images, textual messages and the time and location of when and where the message was posted. Hence, these updated and detailed data can tell us how and when individuals are using urban spaces, how they feel in different places and what their intentions and preferences are (Heppenstall, Malleson, & Crooks, 2016). In contrast to "conventional" or "traditional" social and functional data, collected by survey and census, a DSN provides objective and subjective data on various social and functional activities, travel behaviour and daily space-time movement at different times and spaces, providing great potential for urban research and modelling.

Location-based DSNs, such as Twitter and Foursquare, are particularly important for exploring people's spatial behaviour and for sensing their spatial and temporal preferences in urban locations) Longley, Adnan, & Lansley, 2015; Shelton, Poorthuis, & Zook, 2015). Because of this potential, the respective data are used to capture spatial behaviour and subjective attitudes of people in various urban contexts (e.g. Arribas-Bel, Kourtit, Nijkamp, & Steenbruggen, 2015; Wyly, 2014) and for urban modelling. On the urban scale, Shen and Karimi (2016) used social-media check-in data for modelling the spatial distribution of housing prices. The data were used to trace people's functional spatial and temporal activities and their preferences for different places at the level of individual street segments. As mentioned earlier, Jiang and Miao (2015) used social-media data for modelling the development of urban agglomerations, that is "natural cities", and to provide insight into these cities' structure and dynamics. Social-media check-in data have also been used to extract nationwide interurban movement flows to model spatial interaction and identify regionalization (communities) in China's urban system (Liu, Sui, Kang, & Gao, 2014). In a similar way, Lovelace, Birkin, Cross, and Clarke (2015) examined the potential of social media data (geo-tagged Twitter messages) for modelling movement flows for retail shopping centres by means of spatial interaction models.

DSN data can also be integrated with other sources of "big spatial data", obtained by ICTs, including real-time and volunteered geographic information (Goodchild, 2007). For example, studies have used the smart card databases of individual personal movements through public transportation to investigate mobility patterns in the metropolitan areas of London, Singapore and Beijing (Roth, Kang, Batty, & Barthélemy, 2011; Zhong, Müller, Huang, Batty, & Schmitt, 2014). The integration of DSN data with such large, detailed, movement-flow data sets reveals polycentric structures in metropolitan areas, such as the spatial relationships between employment centres and residential housing.

DSN data also have great potential for enhancing AB modelling, especially movement-flow models (Heppenstall et al., 2016). For example, current AB pedestrian movement models are based on observed aggregate volume data that were collected in selected places, with no information on individual movement paths (i.e. origin, destination and length of the movement path), or people's intentions and feelings during movement. As a result, application of these models in existing and planned urban environments is limited to predicting movement volume distributions (e.g. identifying activity hot spots) throughout the street network, but not movement flows between diverse locations in the street network. However, the rapid development of GPS-based devices (including smartphones) for tracking pedestrian routes and of geographical information technology (e.g. Shoval & Isaacson, 2006) has allowed the capture of pedestrian movement flows in time and space at the level of individual movement paths. These GPS-based data can be integrated with other social media data to capture how people behave and feel in different places. In AB modelling, such subjective data are essential to the definition of behavioural rules regarding agents' interactions with other agents and with urban environments and thus advance our understanding of the pedestrian flow patterns created and the conditions affecting these patterns. When used in this way, DSN data may complement other methods, such as virtual reality, in tracking participant experiences in urban space (Natapov & Fisher-Gewirtzman, 2016).

It follows that geo-referenced social media data may be used for calibrating and validating urban modelling, especially spatial interaction and AB models at the large and local scale, respectively. However, data usability issues have arisen when employing DSN data to represent activity patterns, particularly with regard to sampling (i.e. the data may not reliably represent the true underlying population), privacy, and ethics (i.e. it is not clear if people agree or are aware that their social-media contributions are being used for research and private corporations) and context-related uncertainty (e.g. it is difficult to derive a person's intended meaning from the texts; Boyd & Crawford, 2012; Longley et al., 2015). These usability issues point to the limited application of DSN data so far in urban modelling. Therefore, much more effort should be dedicated to the verification and improvement of DSN data usability, considering their potential.

DSNs as a component of urban modelling

DSNs can enhance urban modelling, not only as a source of data but also as an integral part of that data. This is so because a DSN not only *tells us* how and when individuals use urban spaces; it *affects* how and when individuals use urban spaces. In the following, we discuss the potential implications of a DSN on the formation of activities and movement patterns in the city and, therefore, its relevance for urban modelling.

Herrera-Yagüe et al. (2015) have reported that, on the urban scale, unlike the country scale, geography plays only a minor role in the formation of social networking communities within cities. It is also well known, as suggested by the concepts "network individualism" and "individual-based network" (Neutens, Schwanen, & Witlox, 2011; Wellman, 2002), that contemporary society is in a state of transformation, from a society centred on local geographical environments to one rooted in individuals' interactions as derived from a DSN. That is because a DSN, on the urban scale, allows individuals to contrive social networks with some degree of independence from their geographical environment and to be affected by them, it is reasonable to assume that DSN may impinge on the city's dynamics.

First, the location-based, social-media data concerning activities and services that people share can affect the spatial distribution of those same activities. That is the preferred places of activity are determined according to the interaction between people in the social networks (e.g. recommendations of places to visit made by DSN users) in addition to the interaction between street-network structure and movement flows (i.e. movement-driven activities). The decreasing role of geographical distance and centrality in the formation of land-use patterns implies that DSN may contribute to the *decentralization* of functional and social activities. The same is true of travel behaviour. The growing use of Global Positioning System (GPS)–based navigation systems, such as Waze and Google Maps, may modify travel behaviour preferences, that is the selection of shortest routes and/ or the time to set out. This trend may lead to more dispersed aggregate space-time movement patterns than have been observed previously, reflected in the greater use of shortest metric routes that pass through local and secondary roads, including residential streets (De Baets et al., 2014). Hence, a DSN enhances the formation of a more *equitable urban space* with respect to land-use distribution and movement patterns.

Second, a DSN may also affect the distinctions between urban areas. As aggregative activity patterns in urban space also result from individual-based networks (Neutens et al., 2011), individuals who belong to the same "physical" community or urban area may enjoy different action spaces. Such a situation contains the potential to weaken the boundaries between urban areas and to cause greater overlap between them. Such effects may induce a *semi-lattice structure* of urban environments (Alexander, Ishikawa, & Silverstein, 1977), according to which each part of the city is well connected to its surrounding area and to the city as a whole. Hence, a DSN may intensify globalization and connectivity trends in urban space.

Third, the real-time interactions taking place in social media contribute to the simultaneity of relationships among travel behaviour, space-time movement flows and land-use patterns and thus contribute to accelerating change and greater complexity in cities. Moreover, because a DSN reinforces the individual's ability to influence entire aggregate activity patterns in the city, they may become a core factor of bottom-up urban self-organization.

Concluding remarks: advantages and shortcomings of DSNs

Big data has caused a paradigm shift in data-driven research (Kitchin, 2014; Chen & Zhang, 2018). Within this frame, DSNs are able to provide excellent spatial behaviour information. Moreover, the explosion of new social media linked to the use of ICT and new travel behaviour patterns add a new layer of complexity to the understanding of what a city is. They not only place into question traditional models and definitions of what characterizes a city but also generate new challenges for urban comparative studies. These challenges, however, open the possibility of new reformulations and theoretical considerations and add to the claim that a new understanding of urbanization is needed.

First of all, given the potential implications of DSNs on urban dynamics, and considering their increased involvement in contemporary urban society, DSNs should be considered when constructing urban models. This means that spatial behaviour within urban models may eventually be defined according to the information obtained from DSN, especially in real time (e.g. the selection of activity places, movement destinations, and movement paths). In other words, urban models in future may more accurately and robustly represent people's spatial behaviour in contemporary cities.

Second, DSNs and the new definition of urban areas expand the functional criteria to categorize cities. Furthermore, urban system complexity requires simultaneous consideration of multiple issues, processes and outcomes. DSNs have the potential to establish a dialogue with the other approaches, such as the ones that explore the relations between the strict urbanized space and the other functional spatial scales within the capitalist system.

With all this established, it is also important from a policy perspective to consider the limitations of phenomena such as the one provided by DSN data when helping to reveal activity patterns and to define what a city is and how it can be applied to inform and assist urban policies.

Looking at the source and characteristics of the data, it is necessary to consider that DSN data are generated from many different sources and in different formats (images, audio, tweets, etc.), which are unstructured. These diverse formats and volume of collected geospatial big data pose challenges in storing, managing, processing, analysing, visualizing, and verifying the quality of data (Chen & Zhang, 2018). Quality assessment issues emerge, as data also raises problems of uncertainty and trustworthiness. For instance, Longley et al. (2015), in their study of activity patterns of Twitter users in London (UK), conclude that, although they have gathered a rich digital depiction of individuals, based on participation in social networks, "this falls far short of universal coverage of the population at large, and we have no reliable means of extrapolating from the profiles of self-selecting individuals to the population at large" (pp. 482–483). The authors conclude the problem is essentially that, in these types of data, it is impossible to understand the precise demographics for which we have information. A second issue that they identify as relevantly constraining is that there is no demographic profile of the users in the Twitter user information database. This implies that, without population segments, ultimately the data, regardless of its large volume, only serves to improve our understanding of patterns of movement and flow in themselves.

In focusing on data modelling and structuring, methodological challenges also emerge. First, data-mining software tools reveal limitations on handling large and complex data (Al Nuaimi, Al Neyadi, Mohamed, & Al-Jaroodim, 2015). Second, and most important, although advances in new modelling and methodologies used in social media (see Cranshaw, Schwartz, Hong, & Sadeh, 2012; Chen & Zhang, 2018), as Batty et al. (2012) assert, there is still the need to elaborate new methods to manage the large volume of urban data that provides information across spatial and temporal scales.

From a research perspective, as Kitchin (2013) highlights, one of the main constraints in the use of big data to inform public policy is that big data is mainly generated by privately owned businesses and government. In this scenario, researchers have, paradoxically, highly limited access.

Where access is gained, a number of ethical and security challenges arise. A major challenge in using big data generated through social media and a DSN is the problem of security and privacy issues. Databases may include confidential information. So far, the existence of high-level security policies and mechanisms to protect this data against unauthorized use is questionable.

Also related to ethical and security issues, within the public sector and for policy purposes, information must move over various types of networks and public agencies, which requires high levels of security that cannot always be guaranteed. Furthermore, sharing data and information among different city departments is a challenge considering that agencies and departments have their own information silos (Al Nuaimi et al., 2015).

Another normative aspect to consider is the need to establish priorities. If this is the case and we consider that improving the quality of life for a broad number of urban residents is a priority, as well as providing an answer to global environmental degradation, then we should have an open discussion on which key axes should be considered to establish urban policies, and which role DSNs could play in that.

Finally, urban sociologists, geographers and planners would agree with Jacobs and Munford, who emphasize that the key essence of the city is the formation of social interactions and communities among different people and cultures. That issue must also be explored to determine to what extent DSNs reinforce or constrain this goal.

References

Alexander, C., Ishikawa, S., & Silverstein, M. (1977). *A pattern language: Towns, buildings, construction* (Vol. 2). Oxford, UK; New York, NY: Oxford University Press.

Al Nuaimi, E., Al Neyadi, H., Mohamed, N., & Al-Jaroodim, J. (2015). Applications of big data to smart cities. *Journal of Internet Services and Application, 6*(1), 1–15.

Anas, A., Arnott, R., & Small, K. A. (1998). Urban spatial structure. *Journal of Economic Literature, 36*(3), 1426–1464.

Arribas-Bel, D., Kourtit, K., Nijkamp, P., & Steenbruggen, J. (2015). Cyber cities: Social media as a tool for understanding cities. *Applied Spatial Analysis and Policy, 8*, 231–247.

Batty, M. (2005). *Cities and complexity: Understanding cities with cellular automata, agent-based models, and fractals*. Cambridge, MA: Massachusetts Institute of Technology Press.

Batty, M. (2008). Fifty years of urban modelling: macro statics to micro dynamics. In S. Albeverio, D. Andrey, P. Giordano, & A. Vancheri (Eds.), *The Dynamics of complex urban systems: An interdisciplinary approach* (pp. 1–20). Heidelberg: Physica-Verlag.

Batty, M. (2013). *The new science of cities*. Cambridge, MA: Massachusetts Institute of Technology Press.

Batty, M., Axhausen, K. W., Giannotti, F., Pozdnoukhov, A., Bazzani, A., & Wachowicz, M. (2012). Smart cities of the future. *European Physical Journal Special Topics, 214*, 481–518.

Benenson, I., & Omer, I. (2003). High-resolution census data: A simple way to make them useful. *Data Science Journal, 2*(26), 117–127.

Benenson, I., & Torrens, P. M. (2004). *Geosimulation: Automata-based modeling of urban phenomena*. London: Wiley.

Bettencourt, L. M. A. (2013). The origins of scaling in cities. *Science, 340*, 1438–1441.

Boyd, D., & Crawford, K. (2012). Critical questions for big data: Provocations for a cultural, technological, and scholarly phenomenon. *Information, Communication & Society, 15*(5), 662–679.

Brenner, N. (Ed.). (2014). *Implosions/explosions: Towards a study of planetary urbanization*. Berlin: Jovis Verlag.

Brenner, N., & Schmid, C. (2014). The 'urban age' in question. *International Journal of Urban and Regional Research, 38*(3), 731–755.

Brenner, N., & Schmid, C. (2015). Towards a new epistemology of the urban? *City, 19*(2–3), 151–182.

Burgess, E. (1925). The growth of the city. In R. Park & E. Burgess (Eds.), *The city* (pp. 47–62). Chicago, IL: University of Chicago Press.

Butler, J. (2006/1990). *Gender trouble: feminism and the subversion of identity*. New York, NY: Routledge.

Casellas, A., & Galley, C. (1999). Regional definitions in the European Union: A question of disparities? *Regional Studies, 33*(6), 551–558.

Chen, C. P., & Zhang, C. Y. (2018). Data-intensive applications, challenges, techniques and technologies: A survey on Big Data. *Information Science, 275*, 314–347.

Cranshaw, J., Schwartz, R., Hong, J. I., & Sadeh, N. (2012). The Livehodds Project: Utilizing social media to understand the dynamics of a city. *Proceedings of the Sixth International AAAI Conference of Weblogs and Social Media* (pp. 58–63), Dublin, Ireland.

De Baets, K., Vlassenroot, S., Boussauw, K., Lauwers, D., Allaert, G., & De Maeyer, P. (2014). Route choice and residential environment: Introducing liveability requirements in navigation systems in Flanders. *Journal of Transport Geography, 37*, 19–27.

Delclòs-Alió, X., & Miralles-Guash, C. (2018). Looking at Barcelona through Jane Jacobs's eyes: Mapping the basic conditions for urban vitality in a Mediterranean conurbation. *Land Use Policy, 75*, 505–517.

Desyllas, J., Duxbury, E., Ward, J., & Smith, A. (2003). *Demand modeling of large cities: An applied example from London*. London: Centre for Advanced Spatial Analysis (UCL).

Deuskar, C. (2015). What does urban mean? Retrieved from https://blogs.worldbank.org/sustainablecities/what-does-urban-mean

Dijkstra, L., & Poelman, H. (2012). Cities in Europe: The new OECD-EC definition. *Regional Focus*, *1*, 16. Retrieved from http://ec.europa.eu/regional_policy/sources/docgener/focus/2012_01_city.pdf

Ehlers, E. (2011). City models in theory and practice – A cross-cultural perspective. *Urban Morphology*, *15*(2), 97–119.

Ferguson, P., Fridrisch, E., & Karimi, K. (2012). Origin-destination weighting in agent modelling for pedestrian movement forecasting. *Symposium Proceedings: Eighth International Space Syntax Symposium*, Santiago, Chile.

Fishman, R. (1987). *Bourgeois utopias: The rise and fall of suburbia*. New York, NY: Basic Books.

Florida, R., Gulden, T., & Mellonder, C. (2007). *The rise of the mega region* (p. 31). Toronto: Joseph L. Rotman School of Management, University of Toronto. Retrieved from www.rotman.utoronto.ca/userfiles/prosperity/File/Rise.of.%20the

Frey, W. H., & Zimmer, Z. (2001). Defining the city. In R. Paddison (Ed.), *Handbook of urban studies* (pp. 14–35). London: Sage Publications.

Goodchild, M. F. (2007). Citizens as sensors: The world of volunteered geography. *GeoJournal*, *69*(4), 211–221.

Gottmann, J. (1961). *Megalopolis*. Cambridge, MA: Massachusetts Institute of Technology Press.

Hagerstrand, T. (1967). *Innovation diffusion as a spatial process*. Chicago, IL: University of Chicago Press.

Hall, P. (1988). *Cities of tomorrow: An intellectual history of urban planning and design in the twentieth century*. London: Blackwell Publishers.

Harvey, D. (1989). *The urban experience*. Oxford: Blackwell Publishers.

Harvey, D. (1996). Cities or urbanization? *Cities*, *1*(2), 38–61.

Harvey, D. (2000). *Spaces of hope*. Berkeley, CA: University of California Press.

Heppenstall, A., Malleson, N., & Crooks, A. T. (2016). 'Space, the final frontier': How good are agent-based models at simulating individuals and space in cities? *Systems*, *4*(1), 9, 1–18.

Herrera-Yagüe, C., Schneider, M., Couronné, T., Smoreda, Z., Benito, R., Zufiria, P. J., & González, M. C. (2015). The anatomy of urban social networks and its implications in the searchability problem. *Scientific Reports*, *5*, 11.

Hillier, B., & Penn, A. (1996). Cities as movement economies. *Urban Design International*, *1*(1), 41–60.

Hillier, B., Penn, A., Hanson, J., Grajewski, T., & Xu, J. (1993). Natural movement: Or configuration and attraction in urban pedestrian movement. *Environment and Planning B: Planning and Design*, *20*(1), 29–66.

Howard, E. (1965/1898). *Garden cities of to-morrow*. Cambridge, MA: Massachusetts Institute of Technology Press.

Hoyt, H. (1939). *The structure and growth of residential neighbourhoods in American cities*. Washington, DC: Federal Housing Administration.

Jacobs, J. (1961). *The death and life of great American cities*. New York, NY: Random House.

Jiang, B. (2009). Ranking spaces for predicting human movement in an urban environment. *International Journal of Geographical Information Science*, *23*(7), 823–837.

Jiang, B., & Jia, T. (2011). Agent-based simulation of human movement shaped by the underlying street structure. *International Journal of Geographical Information Science*, *25*(1), 51–64.

Jiang, B., & Miao, Y. (2015). The evolution of natural cities from the perspective of location-based social media. *The Professional Geographer*, *67*, 295–306.

Kitchin, R. (2013). Big data and human geography opportunities, challenges and risks. *Dialogues Human Geography*, *3*(3), 262–267.

Kitchin, R. (2014). Big data, new epistemologies and paradigm shifts. *Big Data & Society*, *1*(1), 1–12.

Le Corbusier. (1987/1933). *The city of to-morrow and its planning*. New York, NY: Dover Architecture.

Lerman, Y., Rofè, Y., & Omer, I. (2014). Using space syntax to model pedestrian movement in urban transportation planning. *Geographical Analysis*, *46*, 392–410.

Liu, Y., Sui, Z., Kang, C., & Gao, Y. (2014). Uncovering patterns of inter-urban trip and spatial interaction from social media check-in data. *PLoS ONE*, *9*(1), e86026.

Longley, P. A., Adnan, M., & Lansley, G. (2015). The geotemporal demographics of twitter usage. *Environment and Planning A: Economy and Space*, *47*(2), 465–484.

Lovelace, R., Birkin, M., Cross, P., & Clarke, M. (2015). From big noise to big data: Toward the verification of large data sets for understanding regional retail flows. *Geographical Analysis*, *48*(2016), 59–81.

Morrill, R. L. (1965). The negro ghetto: Problems and alternatives. *Geographical Review*, *55*, 339–369.

Mumford, L. (1938). *The culture of cities*. London: Secker and Warburg.

Mumford, L. (2000/1937). What is a city? In R. Legates & F. Stout (Eds.), *The city reader* (pp. 92–96). London: Routledge.

Natapov, A., & Fisher-Gewirtzman, D. (2016). Visibility of urban activities and pedestrian routes: An experiment in a virtual environment. *Computers, Environment and Urban Systems*, *58*, 60–70.

Neutens, T., Schwanen, T., & Witlox, F. (2011). The prism of everyday life: Towards a new research agenda for time geography. *Transport Reviews: A Transnational Transdisciplinary Journal*, *31*(1), 25–47.

Omer, I., & Kaplan, N. (2017). Using space syntax and agent-based approaches for modeling pedestrian volume at the urban scale. *Computers, Environment and Urban Systems*, *64*, 57–67.

Omer, I., Rofe, Y., & Lerman, Y. (2015). The impact of planning on pedestrian movement: Contrasting pedestrian movement models in pre-modern and modern neighborhoods in Israel. *International Journal of Geographical Information Science*, *29*, 2121–2142.

Ozbil, A., Peponis, J., & Stone, B. (2011). Understanding the link between street connectivity, land use and pedestrian flows. *Urban Design International*, *16*(2), 125–141.

Poli, C. (2009). *Città Flessibili: Una Rivoluzione Nel Governo Urbano*. Torino: Instar Libri.

Portugali, J. (1999). *Self-organization and the city*. Berlin: Springer.

Portugali, J. (2011). *Complexity, cognition and the city*. Berlin: Springer, Complexity Series.

Roth, C., Kang, S. M., Batty, M., & Barthélemy, M. (2011). Structure of urban movements: Polycentric activity and entangled hierarchical flows. *PLOS One*, *6*(1), e15923.

Sassen, S. (2001). *Global cities: New York, London, Tokyo* (2nd ed., p. 398). Princeton, NJ: Princeton University Press.

Shelton, T., Poorthuis, A., & Zook, M. (2015). Social media and the city: Rethinking urban socio-spatial inequality using user-generated geographic information. *Landscape and Urban Planning*, *142*, 198–211.

Shen, Y., & Karimi, K. (2016). Urban function connectivity: Characterisation of functional urban streets with social media check-in data. *Cities, 55*, 9–21.

Shoval, N., & Isaacson, M. (2006). Application of tracking technologies to the study of pedestrian spatial behavior. *The Professional Geographer, 58*(2), 172–183.

Taylor, P. J. (2004). *World city network: A global urban analysis*. New York, NY: Routledge.

Tönnies, F. (1979/1887). *Gemeinschaft und Gesellschaft: Grundbegriffe der reinen Soziologie*. Darmstadt: Wissenschaftliche Buchgesellschaft.

Torrens, P. M., Nara, A., Li, X., Zhu, H., Griffin, W. A., & Brown, S. B. (2012). An extensible simulation environment and movement metric for testing walking behavior in agent-based models. *Computers, Environment and Urban Systems, 36*, 1–17.

Vaughan, L., Jones, C. E., Griffiths, S., & Haklay, M. (2010). The spatial signature of suburban town centers. *The Journal of Space Syntax, 1*(1), 77–91.

Wellman, B. (2002). Little boxes, glocalization, and networked individualism. In M. Tanabe, P. van den Besselaar, & T. Ishida (Eds.), *Digital cities II: Computational and sociological approaches* (pp. 10–25). Berlin: Springer-Verlag.

White, R., Engelen, G., & Uljee, I. (1997). The use of constrained cellular automata for high-resolution modeling of urban land-use patterns. *Environment and Planning B, 24*, 323–343.

White, R., Engelen, G., & Uljee, I. (2015). *Modeling cities and regions as complex systems: From theory to planning applications*. Cambridge, MA: Massachusetts Institute of Technology Press.

Wilson, A. G. (1970). *Entropy in urban and regional modelling*. London: Pion Press.

Wirth, L. (1938). Urbanism as a way of life. *American Journal of Sociology, 44*(1), 1–24.

Wright, F. L. (1932). *The disappearing city*. New York, NY: W.F. Payson.

Wyly, E. (2014). The new quantitative revolution. *Dialogues in Human Geography, 4*(1), 26–38.

Zhong, C., Müller, S., Huang, X., Batty, M., & Schmitt, G. (2014). Detecting the dynamics of urban structure through spatial network analysis. *International Journal of Geographical Information Science, 28*(11), 2178–2199.

Žižek, S., Milbank, J., & Davis, C. (2009). *The monstrosity of Christ: Paradox or dialectic?* Cambridge, MA: Massachusetts Institute of Technology Press.

11 Revisiting urban models with information and communication technology data? Some examples from Brussels

Arnaud Adam, Gaëtan Montero and Olivier Finance, Ann Verhetsel and Isabelle Thomas

Introduction

Transport and planning analyses mostly start from simple and available geographic urban data, often driven by administrative considerations. The impact of the delineation of the studied area or the size and shape of chosen basic spatial units is almost always set aside, although their impact on the results of transport models is clearly demonstrated (see Jones, Peeters, &Thomas, 2017; Thomas, Jones, Caruso, & Gerber, 2018). This contribution focuses on the impact of the nature of the data used for the partitioning of Brussels city centre. These partitions are a first step in the process of bridging transport analysis and planning through transport demand revealed by diverse dimensions of the city (e.g. social, morphological, economic, etc.).

Partitioning an urban space into groups of places sharing similar properties is common practice for urban geographers and sociologists since the seminal work of the Chicago school (1930s). It is a way to classify elementary spatial units to better understand their similarities and discrepancies and, above all, to better understand intra-urban spatial structures and dynamics within the city. This is usually done in terms of the socio-economic composition of each place but can also be done in terms of interrelationship between georeferenced populations or that of built-up morphologies: each of these aspects are relevant to figure out the intra-urban homo- or heterogeneity. Identifying parts of a city that are similar is part of a nomothetic approach of geography, as opposed to an idiographic one focusing on the strict unicity of each place. Stressing common properties of places avoids putting too much emphasis on too specific results of the same processes. We take part in this approach here by mobilizing new data and recent innovative methods and by applying them to Brussels.

Quantitative revolution (1960s) has facilitated these analyses methodologically, with the development of factorial and cluster analysis, and faster and faster computing systems, even if available data to describe elementary spatial units were rather scarce (urban factorial ecology). Digital revolution has recently exponentially increased the amount of available data (Floridi, 2012; Kitchin, 2013) and, among them, a lot of localized information. This does not mean that these data did not exist before, yet the detection, the registration, the share and the use of these data reached an unprecedented level since the end of the 20th century. The risk with these new emerging data sciences is that some scientists dive into data before elaborating on the specific goals of the research and could lead to data crunching rather than modelling.

Two general kinds of information are commonly used to classify urban elementary spatial units. First, attributes characterizing each elementary spatial unit can be used to compare places based on these specific properties. These attributes can either describe the population living in the units, the economic activities or the characteristics of the built environment. Uni-, bi- and multivariate analyses are used to consider one or many attributes (X_j) to characterize locations, sometimes followed by clustering places into groups sharing common properties (i.e. clusters). Second, geographers use information about the functional interrelationship (w_{ij}) between couples of elementary spatial units i and j. In that case, graph theory can help geographers to analyse the network of interrelations between places, as well as community detection algorithms. The latter aim at delineating groups of places that are highly connected (i.e. communities). For urban geographers interested since a long time in describing and understanding intra-urban structures and dynamics, the emergence of Information and Communication Technology (ICT) data (Global Positioning System, sensors, etc.) offers new ways to get information about places (X_j) and interrelationships between them (w_{ij}). But what do these new data add to urban geography knowledge?

Limiting urban analyses to ICT data is, for sure, not appropriate nowadays: there are no time series, and the spatial representativeness is often questionable. Some parts of the population can be totally missing in the data coverage.[1] Therefore, censuses and surveys are both still relevant for measuring urban complexity. Censuses have the advantage of being well defined, covering almost the entire population and comparable through time. ICT data measure other aspects, often in a very short time lag and at the individual level (see e.g. Longley, Adnan, & Lansley, 2015; Kitchin, 2013; Miller & Goodchild, 2015).

Cities are complex by definition, and the analysis of their complexity includes, among the already mentioned analyses made on people and interactions, the morphology of their built-up components (see e.g. Thomas & Frankhauser, 2013). Therefore, some other analyses will focus on morphology, by using density, fractal dimension and the natural cities method applied to the footprints of the buildings. This will allow the comparison of characteristics of people and their interrelationships through locations with the morphological features of places. In this chapter, we focus on the Brussels Capital Region (noted as BCR, which is an administrative and political entity on its own), that constitutes the dense urban core of the capital city of Belgium.[2] This allows highlighting intra-urban spatial structures.

Four partitions are here elaborated on a selection of data and methods and are presented in the following sections (by order of appearance): a classical "urban factorial ecology", a community detection based on interrelationships revealed by mobile phone calls and from two different ways of measuring the building footprint: a combination of fractal dimension and the density of buildings, and the "natural city" method. Finally, the two last sections present the conclusions and open the discussion.

Urban factorial ecology: a benchmark

Urban factorial ecology was scientifically very fashionable in early quantitative spatial analysis (1960s; see e.g. Berry & Rees, 1969; Hunter, 1972; Johnston, 1978) and had already been applied to Brussels before (see e.g. Dujardin, Selod, & Thomas, 2008). The aim of these methods is to identify groups of places for which inhabitants have similar characteristics (X_i; socio-economic, ethnic, demographic, etc.) by means of a factorial analysis followed by a clustering method (Pruvot & Weber-Klein, 1984). We here use a principal component analysis and a hierarchical cluster analysis in order to identify spatial clusters and further interpret them in terms of classical "urban models".

In the frame of this chapter, several analyses were performed with different sets of variables, with and without a preliminary component analysis. The results in terms of urban spatial structure are a little sensitive to the used method; here, we develop one example. The data used are provided at the scale of the neighbourhood by the Institut Bruxellois des Statistiques et d'Analyses.[3] The selected attributes (Appendix A) are available online as well as a clear definition of each variable, making the data collection simple and reproducible. Among the 145 neighbourhoods of the BCR, 27 are excluded from the analysis because they are not inhabited (mostly correspond to green spaces or industrial zones). The selection of a large number of variables (48) is inspired by the literature on factorial ecology, in order to catch the socio-economic conditions, as well as the demographic and ethnical conditions of each spatial unit.

As expected, variables are highly correlated and can easily be summarized in four components (see Appendix B for the composition of each component). The score of the two first components are rather organized in concentric structures: the first one around the centre (mainly an age gradient), and the second around the European quarter (mainly a "metropolitan activities" gradient). The third component is organized in sectors and is largely based on income and characteristics of the housing stock: large income and large dwellings in the south-east and low income and small dwellings in the north-west. The fourth component reveals a structure in "donut", highlighting places specialized in residential functions at a certain distance from the city centre.

A ward hierarchical cluster analysis based on the scores of the *Principal Component Analysis* (PCA) is then conducted for grouping the neighbourhoods that share similar profiles. Indicators for finding the best number of clusters (Cubic Clustering Criterion, entropy, Dunn, Calinski-Harabasz, shape of dendrogram) show that the optimal number of clusters turns around five to seven. The solution with six clusters is here illustrated (Figure 11.1-a), six being also a number comparable to the number of mobile phone communities that are extracted and presented in the specific section.

Results (Figure 11.1-a) confirm former analyses conducted (see e.g. Vandermotten & Vermoesen, 1995; Thomas & Zenou, 1999): Brussels combines two structures strongly shaping the city: a concentric and a radial one (Hoyt, 1939). The factorial ecology analysis confirms the existence of two crowns around the Central Business District (CBD). Cluster 2 corresponds to the *CBD* with high values of the scores of the second component, little residential. The *first crown* is mainly residential, characterized by a high population density (positive values on Component 4) and divided into two sectors: Cluster 1 (pink) in the south-east which is mainly residential and better-off (high income, large dwellings) and Cluster 3 (green) in the north and west (low income, small housings). The *second crown* is also divided into two parts even if Cluster 6 (orange) and Cluster 4 (purple) are less marked than those of the first crown. Last, Cluster 5 (blue) corresponds to specific urban structures such as large hospitals, sports halls or cultural centres.

Community detection in ICT data: phone basins

Nowadays, there is no need to wait for official census data to "measure" the city: the emergence of new sources of data now allows to sense its "pulses" in real-time. Contrarily to census data describing attributes of places in terms of socio-economic, demographic and ethnic characteristics of their inhabitants (previous section), ICT data enable to characterize and follow people in space and time. ICT data are particularly useful to monitor daily life and short-term processes by detecting and measuring changes within space (Lee & Lee, 2014). They undoubtedly open new perspectives in interaction analyses and urban geography (Batty et al., 2012). This also allows the capture of intra-urban interrelationships between places (w_{ij}). The availability of such data increases every day due to the multiplication of apps on smartphones that enable following people and their geolocations.

We here limit ourselves to one example: mobile phone calls. They already have been proved to be very useful in monitoring and mapping the de facto population, as well as people's spatial mobility and social networks (Blondel, Krings, & Thomas, 2010; Griffiths, Hostert, Gruebner, & Van der Linden, 2010; Batty et al., 2012). It enables one to better grasp the interactions between people and their environment (see e.g. Ahas et al., 2015; Deville et al., 2014). With this type of data, it is common to extract groups of people or locations with remarkably high interactions, leading to what is called communities.

If this kind of data are nowadays very attractive for approximating "social networks", they also appear to have major limitations (see e.g. Calabrese, Ferrari, & Blondel, 2014) including a time-consuming "cleaning" phase. The data are not publicly available, and several operators share the market with no information on market shares. It is quite common that the data are aggregated due to confidentiality issues (no distinction between professional and private use; calls are, in our case, located to the closest antenna and not at the exact location of the call).

We here use one month of phone communications of one of the three major service providers operating in Belgium (April/May 2015). The database includes all mobile phone calls, ignoring their duration, between two phone numbers from the provider, geocoded by antenna at the hour of the call. The data set is here limited to the calls for which both antennas are located in the BCR, that is more than 4.9 million calls. To approximate the coverage of each antenna and to improve the readability of the maps, a Voronoi diagram is designed around each antenna (see Adam et al., 2017).

Methods of community detection based on *modularity* maximization are often adopted to detect communities of nodes that interact in large data sets (Newman, 2004). The inputs of these methods are limited to links between nodes (the weights associated with the edges). If the nodes are geocoded, results can further be mapped, and a spatial partition of the studied area is then obtained. To group together the nodes that are tightly connected and hence detecting communities, these methods search for a compromise between minimizing the connections between nodes classified in different communities (the *cut*) and maximizing the number of communities (the *diversity*; Delvenne, Yaliraki, & Barahona, 2010). The *Louvain Method* is considered as a standard heuristic to maximize the *modularity* value despite its limitations (see e.g. Fortunato, 2009; Traag, 2013; Delvenne, Schaub, Yaliraki, & Barahona, 2013). It is commonly applied because it quickly finds partitions of nodes that maximize the *modularity* without defining a priori central places and/or thresholds (Adam et al., 2018a, Thomas et al, 2017).

Applied to the BCR mobile phone data set, the Louvain Method endogenously detects six communities that can be called "phone basins". A phone basin corresponds to antennas that have a higher propensity to call other antennas than any antenna classified in other communities. In order to avoid a suboptimal solution, the algorithm is run 1,000 times (Adam et al., 2018b). Hence, for each partition, each node can be characterized by the percentage of runs for which it is associated to the same community and so measuring the stability of the partition.

A striking first feature is that most contiguous Voronoi diagrams belong to the same community, this means that calls are more numerous between antennas that are closely located and that people call more often their closest neighbours than those farther located (Tobler, 1970): geography still matters! Table 11.1 represents the matrix of calls emitted and received between the six communities. For each community, the highest number of calls is observed between antennas belonging to the same community; they correspond to the diagonal of the matrix. The intra-community calls represent more than 50% of the calls emitted or received by each community at the exception of Community 6 (45%). In terms of calls made between communities, Table 11.1 and Figure 11.1-c show different realities if the direction of the calls is taken into account. For instance, on one hand, a high number of calls is made from Community 1 to Community 2 (116,986 calls), and on the other, the antennas belonging to Community 2 were many times in communications with antennas from Community 4 (155,949 calls) as well as of Community 1 (123,096 calls).

Table 11.1 Number of calls between communities

Communities	1	2	3	4	5	6
1	684,066	116,986	63,914	67,722	75,337	38,475
2	123,096	741,082	67,416	155,949	54,281	72,337
3	70,716	71,856	263,209	49,667	40,955	35,133
4	71,806	154,892	45,944	629,775	81,803	83,723
5	75,663	52,409	36,826	78,132	316,415	38,700
6	41,577	72,538	33,734	85,011	40,846	217,392

Figure 11.1-b clearly reveals the spatial organization of calls in the BCR. Two of them are located in the middle of the study area: a community centred on the Pentagon[4] with an extension to the North Railway Station (Community 4) and another one around the EU offices (Community 1). These two communities concentrate 43% of the calls made within the BCR; the other communities are further organized in sectors around these two.

Morphometrics of built-up footprint

The objective of this section is to give an image of the built-up disparities and further partition the city in terms of built-up urban similar morphologies. The literature mentions many indices mainly issued from landscape ecology for measuring and characterizing urban morphologies (see e.g. Medda, Nijkamp, & Rietveld, 1998; Schwarz, 2010; Caruso, Hilal, & Thomas, 2017). We here select and compare two methods to quantitatively grasp the morphological reality of Brussels: fractal dimension combined with density and the "natural cities" (presented in the following sections).

We use the 2009 Cadastral footprint of the buildings as well as the centroids of each building. Every isolated small (less than 20 m²) building was erased from the database. Analyses are performed on an area larger than the BCR (an area encompassing the entire former province of Brabant that surrounds the BCR) in order to avoid border effects. From this large set, we here isolated and present the results of the BCR. In this analysis, we take care of these border effects because it is well known that fractal dimension and "natural cities" methods are particularly sensitive to these border effects (Montero, Tannier, & Thomas, 2018).

Fractal dimension and density

Built-up urban fabrics with complex geometrical features cannot be described only by simple tools based on Euclidean geometry. Fractal geometry provides an interesting alternative to compare irregular forms, even at different spatial scales (Batty & Kim, 1992). Fractal dimension (D) characterizes the scaling behaviour of fractals, that is the fact that the same structure statically appears on smaller nested scales. For surfaces, D varies between 0 and 2: 2 corresponds to a homogeneous pattern where the mass is distributed uniformly over space, while 0 is a quite unrealistic value corresponding to isolated points without any particular spatial arrangement. The more D is different from 2, the more the patterns show empty areas of different sizes; these empty areas are distributed according to a strong hierarchical law. When $D < 1$, the pattern consists of disconnected, isolated elements concentrated in clusters which are separated by lacunas of different sizes. Hence, the lower D is, the less homogenously the built-up areas are distributed over space. Fractal dimension is independent of the unit of measurement (see e.g. Thomas, Frankhauser, & De Keersmaecker, 2007; Thomas, Cotteels, Jones, & Peeters, 2012).

Fractal dimension can be seen as a proxy of mean density, but the two are not spatially equivalent. Fractal dimension relates to morphology (the internal structure of the built-up areas) while mean density gives a rough idea of the occupation of the area. Geographically weighted fractal analysis is used for computing local fractal dimensions. This method mixes the sandbox multifractal algorithm (see e.g. Vicsek, 2002) and a geographically weighted regression with a kernel to estimate the fractal dimension of cells in a regular grid (Sémécurbe, Tannier, & Roux, 2019). In order to provide results at the finest possible resolution and to avoid estimation problems of the fractal dimension, we use 250-m × 250-m resolution cells.

Density is here simply expressed as the ratio of the surface occupied by the buildings divided by the total surface of the basic spatial unit. The studied area is hence covered by a grid of squared cells for which two values are calculated: density and fractal dimension. A Ward clustering analysis is applied. Results are reported in Figure 11.1-d. The data set is clustered in four crowns in a clear centre–periphery structure. From the outskirts to the centre: (1) low density and non-homogeneous built-up surfaces, (2) average density and average fractal dimension, (3) high density and average homogeneity and (4) very high density and high homogeneity. This structure reminds clearly the classical concentric urban structure model of Alonso-Muth (see e.g. Verhetsel, Thomas, & Beelen, 2010).

From "natural cities" to "natural urban clusters"

A set of papers has recently been initiated by Jiang around the concept of "natural cities" (see e.g. Jia & Jiang, 2011; Jiang, 2013; Jiang & Miao, 2015). The methodology relies on a head/tail division rule to derive "natural" clusters (called "cities" in Jiang's papers), based on the assumption that there are far more smalls things than larger ones. A triangulated irregular network (TIN) is used, made up of individual locations that are considered as *nodes* (initially street nodes and later any location-based social media users such as Twitter or Brightkite, here we adapt the method by using the centroid of each building in order to make the results comparable with the previous section). The method identifies the edges between the nodes (on the TIN) smaller than the average length of the edges as limits of "natural cities". The method relies on the fact that the length of the edges follows a hierarchical distribution where the mean length is used as a threshold for delineating the "cities" (large number of small edges corresponding to cities and small number of large edges corresponding to rural areas). The method appears to be very attractive in a "mechanical" way, but questions remain about its anchoring in urban geography theory and its combination with urban functional issues (see Montero et al., 2018, for a critical and comparative analysis). We here decided to use it for detecting intra-urban clusters within the BCR: Can the BCR be partitioned into "*natural urban clusters*"? Are some parts of the city characterized by tighter networks than others?

If several authors have already supported the use of streets networks for delineating cities (Jia & Jiang, 2011; Arcaute et al., 2015; Masucci, Arcaute, Hatna, Stanilov, & Batty, 2015), Thomas et al. (2012) showed that the spatial organization of buildings within a city is rather different from that of street networks. Hence, in order to compare results with the previous section, we here consider the centroid of each building as the nodes used in the TIN.

The method delineates zones where the spatial proximity of the centroids is higher than the mean length in the study area (48 m). Patches are clearly identified and mapped (Figure 11.1-e). The white space between the patches corresponds to large boulevards, wide infrastructures as railways, green spaces, squares, etc., or simply built-up wards with larger distances between their centroids (larger buildings or larger gardens). Some coloured patches are large and reflect the history and geography of the city (see e.g. Vandermotten & Vermoesen, 1995). Some others are very small, corresponding to specific allotments or pinpoint urban projects. Figure 11.1-e clearly tells another story even if some resemblance exist; there is a clear resemblance between socio-economic realities (Figure 11.1-a) and built-up landscapes (Figure 11.1-e) but the method used for extracting urban built-up landscapes influences the result (Figure 11.1-d and Figure 11.1-e are clearly different).

Complementarity of the four partitions

The four partitions presented in Figure 11.1 result from a unique objective (understanding the city), but the type of data and the methodology, as well as the basic spatial units, differ. Yet some commonalities are observed between the partitions, with some parts of the city being for example encompassed both in a given cluster in the factorial ecology and a specific community into the group of antennas.

An attempt at synthesis is proposed in Table 11.2. The spatial structure of each partition is clearly not similar, at the exception of the specificity of the core area that we found in each of them. Communities centred on the Pentagon and on EU offices are grouped in a socio-economic cluster characterized by a concentration of offices and urban facilities (CBD). The CBD and the first-crown in the factorial ecology are characterized by a dense and homogeneous organization of the built-up environment according to the morphological classification. On the opposite, it is obvious that the comparison of a pure centre–periphery gradient (clustering using the fractal dimension and the built-up density) and a sectoral structure (communities based on cell phone data) will not lead to a strong resemblance between the two partitions.

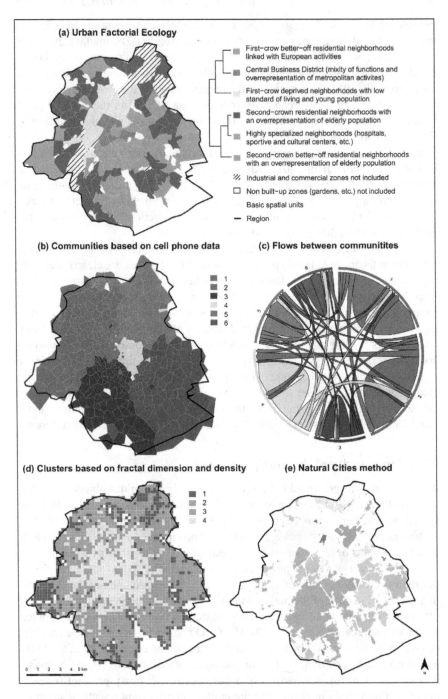

Figure 11.1 Mapping Brussels differently (coloured version of this figure is to be found in https://atlas.brussels/publications/chapter-revisiting-urban-models-coloured-versions-of-figures/)

Table 11.2 Comparison of the four partitions of Brussels

	Type of data	*Basic spatial units*	*Approximative spatial structure of the clustering*
Factorial ecology (PCA + clustering)	Socio-economic, conventional (census)	Statistical neighbourhoods	Combined centre– periphery and sectoral structures
Communities based in cell phone	Relational, unconventional	Voronoi diagrams around antennas	Sectoral structure
Fractal dimension and density	Built-up footprints	Cells grid	Centre–periphery structure
"Natural cities"	Built-up footprints (centroid)	/	Multiple nuclei

The main explanation of the difficulties in comparing the different partitions origins from the fact that the elementary spatial unit is not the same in each analysis. The strict comparison is probably biased by the difference of size, shape and distribution of the basic spatial units (statistical neighbourhoods, Voronoi cells, grid cells) in the city. This is a constraint which we cannot overcome easily because the basic spatial unit is given by the data themselves, and this has to be kept in mind when comparing the different results. It reduces the possibilities of integrating the different kinds of data in a single approach.

Conclusion

With the example of Brussels, this contribution confirms the complexity of the urban structure and the difficulties in fitting data, method and objective. There is no unique way of measuring this complexity: each database and methodology leads to different spatial structures. With this case study, the classical urban models are partly rediscovered by means of new tools and new data. Concentric model (Alonso-Muth/Model of Burgess), sectoral model (Hoyt) and multiple nuclei model (Harris and Ullman) are clearly appearing in Figure 11.1 confirming former studies: Brussels is a superposition of different spatial patterns. Also, for results based on mobile phone data, there is a strong contiguity in the communities: one phones more people who are located closer, reminding one of Toblers's first law of geography.[5]

The definition of the data needs to be controlled, as well as the underlying processes and the research objectives. We cannot let the data speak by themselves, and this is especially true for new ICT data sources. We need to capture the meaning of the data. For ages, geographers have struggled for individual locations to understand processes; ICT data do not change this, as most data are protected by confidentiality reasons (and further, once again, aggregated here by antennas).

Because we did not have the opportunity to conduct all four analyses with the same basic spatial units, the partitions obtained in this contribution are simply juxtaposed and compared between them. Future works will now have to be dedicated to the combination of partitions.

Finally, visualizing data is now current for ICT data, but modelling and linking the results to theory and planning are challenging exercises. There is a clear need to develop and understand methods and link them, first, with spatial theories in a multidisciplinary context before the results can be applied in planning exercises. ICT data are an opportunity to renew quantitative geography, especially as big data enables managing the complexity of interrelationships at local/global scales to add information to conventional data, but big data do not replace them. There is further a clear need for analysis comparing the obtained networks at different scales and for different node definitions (assortativity). Each of the three dimensions we focused on in this chapter brings a distinct point of view and a specific added value. One of the further steps could now be to use methods that mix community detection based on the interactions and cluster analyses based on similarities between places. ICT data are not only a smokescreen: they open new avenues for further dynamic multidisciplinary analyses and for modelling urban realities for the purpose of transport planning.

Acknowledgements – This article was partly written in the framework of the Bru-Net research project financed by Innoviris (Brussels). The authors also wish to thank the telephone provider for the excellent collaboration in the creation of a telephone database and for agreeing to this publication.

Notes

1 For example, there is an over representation of young and male Twitter users in the total resident population in London (Longley et al., 2015).
2 Results shown here are parts of a broader project named Bru-Net that is financed by the Brussels Research Agency and that aims at measuring spatial communities in the whole metropolitan area around Brussels (former province of Brabant). We here limit our analyses to the Brussels Capital Region (BCR), the 19 administrative municipalities of the Brussels core.
3 The "Monitoring des Quartiers" is an interactive tool available online. It has the objective to make available a selection of indicators that characterize disparities and dynamics within the BCR. Not all data are from the same date. Its availability makes the analysis easily reproducible. Retrieved from http://ibsa.brussels/chiffres/chiffres-par-quartier#. WnoZLOjOXyQ.
4 Usual name of the area located inside the former 14th-century walls of the city.
5 "Everything is related to everything else, but near things are more related than distant things" (1970).

References

Adam, A., Charlier, J., Debuisson, M., Duprez, J.-P., Reginster, I., & Thomas, I. (2018a). Bassins résidentiels en Belgique: deux methodes, une réalité ? *L'Espace Géographique*, *1*, 35–50.

Adam, A., Delvenne, J.-C., & Thomas, I. (2017). Cartography of interaction fields in and around Brussels: Commuting, moves and telephone calls. *Brussels Studies*, *118*. Online since December 18, 2017.

Adam, A., Delvenne, J.-C., & Thomas, I. (2018b). On the robustness of the Louvain method to detect communities: Examples on Brussels. *Journal of Geographical Systems, 20*(4), 363–386.

Ahas, R., Aasa, A., Yuan, Y., Raubal, M., Smoreda, Z., Liu, Y., Zook, M. (2015). Everyday space – time geographies: Using mobile phone-based sensor data to monitor urban activity in Harbin, Paris, and Tallinn. *International Journal of Geographical Information Science, 29*, 2017–2039.

Arcaute, E., Hatna, E., Ferguson, P., Youn, H., Johansson, A., & Batty, M. (2015). Constructing cities, deconstructing scaling laws. *Journal of Royal Society Interface, 12*, 20140745. doi:10.1098/rsif.2014.0745

Batty, M., Axhausen, K., Giannotti, F., Posdnouktov, A., Bazzani, A., Wachowicz, M. . . . Portugali, Y. (2012). Smart cities of the future. *The European Physical Journal Special Topics, 214*, 481–518. doi:10.1140/epjst/e2012–01703–3

Batty, M., & Kim, S. (1992). Form follows function: Reformulating urban population density functions. *Urban Studies, 29*, 1043–1070.

Berry, B., & Rees, P. (1969). The factorial ecology of Calcutta. *American Journal of Sociology, 74*, 445–491.

Blondel, V., Krings, G., & Thomas, I. (2010). Regions and borders of mobile telephony in Belgium and in the Brussels metropolitan zone. *Brussels Studies.* https://doi.org/10.4000/brussels.806

Calabrese, F., Ferrari, L., & Blondel, V. (2014). Urban Sensing using mobile phone network data: A survey of research. *ACM Computing Surveys, 47*, 1–20.

Caruso, G., Hilal, M., & Thomas, I. (2017). Measuring urban forms from inter-building distances: Combining MST graphs with a Local Index of Spatial Association. *Landscape and Urban Planning, 163*, 80–89.

Delvenne, J.-C., Schaub, M. T., Yaliraki, S. N., & Barahona, M. (2013). The stability of a graph partition: A dynamics-based framework for community detection. In *Dynamics on and of complex networks* (Vol. 2, pp. 221–242). New York, NY: Springer.

Delvenne, J.-C., Yaliraki, S. N., & Barahona, M. (2010). Stability of graph communities across time scales. *PNAS (Proceedings of the National Academy of Sciences of the USA), 107*, 12755–12760. https://doi.org/10.1073/pnas.0903215107

Deville, P., Linard, C., Martin, S., Gilbert, M., Stevens, F., Gaughan, A. . . . Tatem, A. (2014). Dynamic population mapping using mobile phone data. *PNAS (Proceedings of the National Academy of Sciences of the USA), 11*(45), 15888–15893. https://doi.org/10.1073/pnas.1408439111

Dujardin, C., Selod, H., & Thomas, I. (2008). Residential segregation and unemployment: The case of Brussels. *Urban Studies, 45*(1), 89–113.

Floridi, L. (2012). Big data and their epistemological challenge. *Philosophy & Technology, 25*, 435–437. https://doi.org/10.1007/s13347-012-0093-4

Fortunato, S. (2009). Community detection in graphs. *Physics Reports, 486*, 75–174. https://doi.org/10.1016/j.physrep.2009.11.002

Griffiths, P., Hostert, P., Gruebner, O., & Van der Linden, S. (2010). Mapping megacity growth with multi-sensor data. *Remote Sensing of Environment, 114* (2), 426–439. doi:10.1016/j.rse.2009.09.012

Hoyt, H. (1939). *The structure and growth of residential neighborhoods in American cities.* Washington, DC: Government Printing Office.

Hunter, A. (1972). Factorial ecology: A critique and some suggestions. *Demography, 9*(1), 107–117.

Jia, T., & Jiang, B. (2011). Measuring urban sprawl based on massive street nodes and the novel concept of natural cities. Preprint, arxiv.org/abs/1010.0541

Jiang, B. (2013). Head/tail breaks: A new classification scheme for data with a heavy-tailed distribution. *The Professional Geographer*, *65*(3), 482–494.

Jiang, B., & Miao, Y. (2015). The evolution of natural cities from the perspective of location-based social media. *The Professional Geographer*, *67*(2), 95–306.

Johnston, R. (1978). Residential area characteristics: Research methods for identifying urban sub-areas analysis and factorial ecology. In D. T. Herbert & R. J. Johnston (Eds.), *Social areas in cities: Processes, patterns and problems* (pp. 175–217). Chichester, West Sussex: John Wiley and Sons.

Jones, J., Peeters, D., & Thomas, I. (2017). Scale effect in a LUTI model of Brussels: Challenges for policy evaluation. *European Journal of Transport and Infrastructure Research (EJTIR)*, *17*(1), 103–131.

Kitchin, R. (2013). Big data and human geography: Opportunities, challenges and risks. *Dialogues in Human Geography*, *3*(3), 262–267.

Lee, J., & Lee, H. (2014). Developing and validating a citizen-centric typology for smart city services. *Government Information Quarterly, ICEGOV*, *2012*(Supplement 31), S93–S105. https://doi.org/10.1016/j.giq.2014.01.010

Longley, P. A., Adnan, M., & Lansley, G. (2015). The geotemporal demographics of twitter usage. *Environment and Planning A: Economy and Space*, *47*, 465–484. https://doi.org/10.1068/a130122p

Masucci, A. P., Arcaute, E., Hatna, E., Stanilov, K., & Batty, M. (2015). On the problem of boundaries and scaling for urban street networks. *Journal of the Royal Society Interface*, *12*(102). http://dx.doi.org/10.1098/rsif.2015.0763

Medda, F., Nijkamp, P., & Rietveld, P. (1998). Recognition and classification of urban shapes. *Geographical Analysis*, *30*(4), 304–314.

Miller, H., & Goodchild, M. (2015). Data-driven geography. *GeoJournal*, *80*(4), 449–461.

Montero, G., Tannier, C., & Thomas, I. (2018). Morphological delineation of an urban space: Critical analysis of three methodologies. *The case study of Brussels* (on going).

Newman, M. E. J. (2004). Detecting community structure in networks. *The European Physical Journal B*, *38*, 321–330. https://doi.org/10.1140/epjb/e2004-00124-y

Pruvot, M., & Weber-Klein, C. (1984). Ecologie urbaine factorielle: essai méthodologique et application à Strasbourg. *L'Espace Géographique*, *13*(2), 136–150.

Schwarz, N. (2010). Urban form revisited – Selecting indicators for characterising European cities. *Landscape and Urban Planning*, *96*(1), 29–47.

Sémécurbe, F., Tannier, C., & Roux, S. G. (2019). Applying two fractal methods to characterise the local and global deviations from scale invariance of built patterns throughout mainland. *Journal of Geographical Systems*, *21*, 271. https://doi.org/10.1007/s10109-018-0286-1

Thomas, I., Adam, A., & Verhetsel, A. (2017). Migration and commuting interaction fields: A new geography with a community detection algorithm? Belgeo [En ligne], 4, doi: 10.4000/belgeo.20507Thomas, I., Cotteels, C., Jones, J., & Peeters, D. (2012). Revisiting the extension of the Brussels urban agglomeration: New methods, new data, new results? *E-Belgeo* (on-line). Retrieved from http://belgeo.revues.org/6074

Thomas, I., & Frankhauser, P. (2013). Fractal dimensions of the built-up footprint: Buildings versus roads: Fractal evidence from Antwerp (Belgium). *Environment and Planning B: Planning and Design*, *40*(2), 310–329.

Thomas, I., Frankhauser, P., & De Keersmaecker, M.-L. (2007). Fractal dimension versus density of the built-up surfaces in the periphery of Brussels. *Papers in Regional Science*, *86*(2), 287–307.

Thomas, I., Jones, J., Caruso, G., & Gerber, P. (2018). City delineation in LUTI models: Review and tests. *Transportation Reviews*, *38*(1), 6–32.

Thomas, I., & Zenou, Y. (1999). *Ségrégation urbaine et discrimination sur le marché du travail: le cas de Bruxelles*. In M. Catin, J.-Y. Lesueur, & Y. Zenou (Eds.), *Emploi, concurrence et concentration spatiales* (pp. 105–127). Paris: Economica.

Tobler, W. (1970). A computer movie simulating urban growth in the Detroit region. *Economic Geography*, *46*, 234–240.

Traag, V. (2013). *Algorithms and Dynamical models for communities and reputation in social networks*. Ph.D. at EPL, Université catholique de Louvain, Louvain la Neuve.

Vandermotten, C., & Vermoesen, F. (1995). Structures sociales comparées de l'espace de trois villes européennes: Paris, Bruxelles, Amsterdam. *Espace Populations Sociétés*, *3*, 395–404.

Verhetsel, A., Thomas, I., & Beelen, M. (2010). Commuting in Belgian metropolitan areas: The power of the Alonso-Muth model. *Journal of Transport and Land Use*, 2(3-4), 109–131.

Vicsek, T. (2002). Complexity: The bigger picture. *Nature*, *418*, 131.

Appendix A

48 variables used in the factorial ecology

% children in the neighbourhood and around, enrolled in kindergartens in the neighbourhood.

% children in the neighbourhood and around, enrolled in a primary school in the neighbourhood.

Population density (inhabitants/sq km).

Offices density (sq meters/sq km).

% of buildings with 5 floors and more.

% of impervious surfaces.

% dwellings built before 1961.

% households unsatisfied with the cleanness of the environment around their housing.

% of population aged 18–29.

% of population aged 65–79.

% population aged 80 and more.

Average age (years).

Mobility index (sum of the immigrants and emigrants divided by total population). Sedentariness index (non-migrants divided by the total population).

% of couples with children in total number of private households.

% of EU population (EU15) in the total population.

% of North Africa population in the total population.

% of Latin America population in the total population.

% of Sub-Saharan Africa population in the total population.

% of foreigners in the total number of inhabitants.

% of inhabitants with French nationality.

% people living alone within the 18–29y old.

% people living alone within the 65 and more.

Mean size of private household.

Index of masculinity (number of men *100 divided by the number of women).

% of the 0–17y old. in the total population.

% population aged 65 and more.Economic dependence coefficient (number of 0–17 y and 65+ y divided by the number of 18–64 y).

Ageing coefficient (%) (population of 65+y divided by the population of 0–17y).

Activity rate (%) (number of actives divided by population of 18–64 y).

Application rate (%) (total number of unemployment people divided by population of 18–64 y)

Unemployment rate (%) (total number of unemployment people divided by the active population)

Medium income of the tax report (€).

% of the salaried within the labor force.Employment rate (%).

Absolute difference between male and female activity rates.

% households residing in an apartment.

% dwellings under 55 sq meters.

% dwellings larger than 104 sq meters.

Average area of a dwelling (sq m).

Average area of a dwelling by inhabitant.

% of the social housing (number of social housing for 100 households).

% dwellings occupied by the owner.

Number of renovation subsidies (for 1000 households) (‰).

% of inhabitants living close from a public transportation stop (250 m from the bus, 400m from a tram and 500m from the metro).

Density of private households (number of private households divided by the area of the district)

% of children (less than 18y) living in a household without labour income.

% inhabitants estimating themselves not to be in good health.

Appendix B

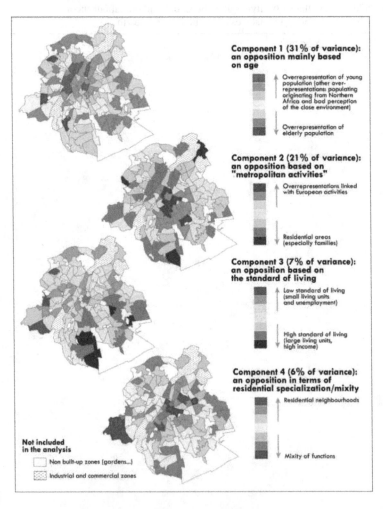

Figure 11.B Four first principal components of the PCA based of variables listed in Appendix A (coloured version of this figure is to be found in https://atlas.brussels/publications/chapter-revisiting-urban-models-coloured-versions-of-figures/)

12 The evolution of natural cities from the perspective of location-based social media

Bin Jiang and Yufan Miao

1. Introduction

Once upon a time, there were no cities, only scattered villages. Over time, cities gradually emerged through the interaction of people or residents; similarly, large or megacities evolve through the interaction of cities or people. This is a thought experiment mentioned by Jiang (2015b), in which he argued that many geographic phenomena, such as urban growth, are essentially unpredictable. Many models in the literature that claim to be able to predict urban growth are in effect for short-term prediction, such as the weather forecast; weather forecasting beyond five days is essentially unforecastable (Bak, 1996). A typical city may have hundreds of years of history, making it nearly impossible to track its growth quantitatively because of a lack of related data. More important, a city grows within a network of cities; one cannot understand a city's growth without considering other related cities. In this paper, we illustrate that emerging social media provide an unprecedented data source for studying the evolution of natural cities (cf. Section 2 for the definition) and subsequently for better understanding structure and dynamics of real cities. Location-based social media, sometimes termed as location-based social networks, such as Flickr, Twitter and Foursquare (Traynor & Curran, 2012; Zheng & Zhou, 2011), refer to a set of Internet-based applications founded on Web 2.0 technologies and ideologies that allow users to create and exchange user-generated content. Location-based social media can act as a proxy of real cities (or human settlements in general) and provide a better understanding of their underlying structure and dynamics.

Not a long ago, there was no social media, only scattered home pages and bulletin board systems created and maintained by individuals and institutions (Boyd & Ellison, 2008; Kaplan & Haenlein, 2010). In the era of Web 1.0, geographic locations were not an issue. However, with Web 2.0, geographic locations have been becoming an important feature of social media. Almost all social media have users' geographic locations enabled, often at the level of meters, when sharing and exchanging user-generated content. Location-based social media enables users to track individual historical trajectories, their friends and even the growth of social media. Unlike with conventional cities, the trajectories of social media are well documented by the hosting companies, and unlike conventional census data, social media data are defined at the individual level, often at very fine spatial and temporal scales. Data can be obtained using crawling techniques or through social media's officially released application programming interfaces. This study aimed to showcase how social media's timestamped location data can be utilized to study the evolution of natural cities, thus providing new insights into the underlying structure and dynamics of real cities.

The contribution of this paper can be seen from the three aspects: data, methods, and new insights. This study produced a large amount of data regarding natural cities from the former social media platform Brightkite during its entire 31-month life span. The resulting data have significant value for further study of city growth and the allometric relationship between populations and physical extents (data available here: https://sites.google.com/site/naturalcitiesdata/). We drew on a set of fractal- or scaling-oriented methods to characterize natural cities. These unique methods help create new insights into the evolution of natural cities as well as that of real cities. For example, natural cities demonstrate a striking non-linear property, spatially and temporally (see Section 4). Moreover, the evolution of natural cities can provide a better understanding of social media from a unique geospatial perspective.

This study provides new perspectives, as well as different ways of thinking, to the study of cities and city growth in the era of big data (Mayer-Schonberger & Cukier, 2013). We did not adopt conventional census data, but rather the emerging georeferenced social media data; we did not adopt conventional geographic units or boundaries that are imposed from the top down by authorities, but rather the naturally or objectively defined concept of natural cities, to avoid statistical biases out of the modifiable areal unit problem (Openshaw, 1984); and we did not rely on standard and spatial statistics with a well-defined mean to characterize spatial heterogeneity, but rather power-law-based statistics, driven by fractal and scaling thinking (Jiang 2013, 2015a). Therefore, the underlying ways of thinking adopted in this study are (1) bottom-up rather than top-down, in terms of data and methods, and (2) non-linear rather than linear, or fractal rather than Euclidean in terms of the power-law statistics.

The remainder of this chapter is structured as follows: Section 2 presents the methods in which we define the concept of natural cities and discuss ways of characterizing natural cities. Section 3 presents the data on a monthly basis and shows basic statistics of the data. Section 4 discusses the results and major findings, while Section 5 on the implications of the study. Finally, Section 6 draws a conclusion and points to future work.

2. Methods

In this section, we illustrate and define the concept of natural cities and present various ways of characterizing natural cities. We also discuss how natural cities differ from conventional cities and why the notion of natural cities represents a new way of thinking for geospatial analysis.

2.1 Defining natural cities

To approach the difficult task of defining and describing natural cities, we start with definitions of conventional cities and try to clarify why the conventional definitions are little natural. A city is a relatively large and permanent human settlement. But how large a settlement must be to qualify as a city is unclear. For example, a city in Sweden may not qualify as a city in China. Also, many cities have a particular administrative, legal, and historical status according to their local laws. In the United States, for example, cities can refer to incorporated places, urban areas or metropolitan areas with a sufficient population of, say, at least 10,000. This population threshold can be very subjective and is dependent on the country. This subjectivity is also demonstrated in the physical boundaries of cities, which are legally and administratively determined. Remotely sensed imagery provides new means to delineate city boundaries, but how does one choose an appropriate pixel value as a cut-off for the delineation? Because of these subjectivities, conventional definitions of cities are little natural. How, then, can we define a city in natural ways?

We present three examples of natural cities before formally define the concept. In the first example, natural cities are derived from massive street nodes, including both junctions and street ends. Given all street nodes of an entire country, we can run an iterative clustering algorithm to determine whether a node is within the neighbour of another node. For example, set a radius of 700 m and continuously draw a circle around each node to determine whether any other node is within its circle. This progressive and exhaustive process results in many natural cities; see Figure 12.1a for an illustrative example. In their study, Jiang and Jia (2011) found that millions of natural cities could be derived from dozens of millions of street nodes in the United States using OpenStreetMap (OSM) data (Bennett, 2010). Instead of massive street nodes, the second example relies on a massive number of street blocks to extract natural cities. Jiang and Liu (2012) adopted the three largest European countries: France, Germany, and the United Kingdom for their case studies, again using OSM data. The idea is illustrated in Figure 12.1b in which small blocks (smaller than a mean) constitute a natural city. Although this method sounds very simple, the computation is very intensive for each country, involving millions of street blocks. The third example comes from Jiang and Yin (2014), in which the authors relied on nighttime imagery to derive natural cities. The authors took all pixel values (millions of pixels each valued between 0 and 63) of an image in the United States and computed an average or mean. The mean split all the pixels into two: those greater than the mean and those less than the mean. For the pixels greater than the mean, a second mean was obtained, and it can be a meaningful cut-off for delineating natural cities.

These examples of deriving natural cities point out the importance of the mean's effect, which is based on the head/tail division rule: *Given a variable* X, *if its values* x *follow a heavy-tailed distribution, then the mean (m) of the values can divide all the values into two parts: a high percentage in the tail and a low percentage in the head* (Jiang & Liu, 2012). The head/tail division rule

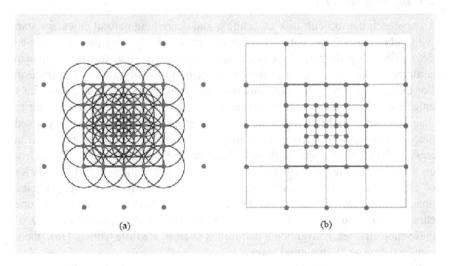

Figure 12.1 (Colour online) Natural cities based on (a) street nodes and (b) street blocks

Source: Jia and Jiang (2010).

Note: Blue rectangles are the boundaries of the natural cities, which are composed of high-density nodes or small street blocks based on the head/tail breaks (Jiang, 2013, 2015a).

has been further developed to a new classification and visualization tool called head/tail breaks for data with the heavy-tailed distribution (Jiang, 2013, 2015a). The heavy-tailed distribution refers to the statistical distributions that are right-skewed, for example, power law, lognormal, and exponential. Obviously, the density of street nodes, the size of street blocks, and the nighttime imagery pixel values all exhibit a heavy-tailed distribution, which implies that there are far more smalls than larges. In this chapter, we build up a huge triangular irregular network (TIN) based on social media users' locations and then categorize these small triangles (smaller than a mean) as natural cities (Figure 12.2); refer to the Appendix for a short tutorial. Section 5 further discusses why the head/tail breaks works so well in delineating natural cities.

Based on these examples, a formal definition of natural cities can be derived. Natural cities refer to human settlements or human activities in general on Earth's surface that are objectively or naturally defined and delineated from massive geographic information of various kinds and based on the head/tail breaks. Unlike conventional cities, natural cities do not need to meet a minimum population requirement. A one-person settlement may constitute a natural city, or even zero people, if natural cities are defined not according to human population but to something else. For example, when natural cities are defined according to street nodes, a natural city derived from one street node may have no people there at all. The reader may question whether this definition makes sense. The definition makes good sense because it provides a new perspective for geospatial analysis

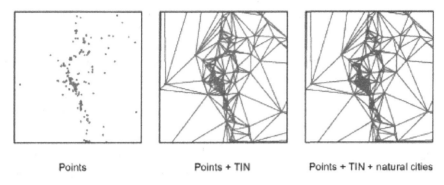

Points   Points + TIN   Points + TIN + natural cities

Figure 12.2 (Colour online) Procedure of generating natural cities (red patches) from points through TIN

and helps us develop new insights into geographic forms and processes (see Sections 4 and 5). That is also the reason that we use the term *natural cities* to refer to human settlements or human activities in general on the Earth's surface. With the concept of natural cities, we abandon the top-down imposed geographic units or boundaries such as states, counties and cities, in order to study geographic forms and processes more scientifically.

2.2 Characterizing natural cities

The rank-size distribution of cities in a region can be well characterized by Zipf's law, that is an inverse power relationship between city rank (r) and city size (N), $N = r^{-1}$ (Zipf, 1949). Simply put, when ranking all cities in decreasing order for a given country, the largest city is twice as big as the second largest, three times as big as the third largest and so on. In other words, a city's size by population is inversely proportional to its rank. Such a simple and neat law is found to hold remarkably well for almost all countries or regions (e.g. Berry & Okulicz-Kozaryn, 2011), although some researchers have challenged its universality (e.g. Benguigui & Blumenfeld-Leiberthal, 2011). Essentially, Zipf's law indicates two aspects: (1) a power-law relationship between rank and size and (2) the Zipf's exponent of one. Most previous studies have confirmed the first aspect but not the second; the Zipf's exponent was found to deviate from one. In other words, the first aspect is not as controversial as the second aspect. Some researchers argued that Zipf's law was primarily used for characterizing large cities rather than all cities. In this study, we chose large natural cities (larger than a mean) to examine whether they followed Zipf's law. The scaling patterns of far more small cities than large ones underlie Zipf's law – a majority of small cities while a minority of large cities. More important, the scaling pattern occurs not just once but recurs multiple times. This is the basis of head/tail breaks (Jiang, 2013), a novel classification scheme for data with a heavy-tailed distribution. In what follows, we illustrate head/tail breaks with a working example.

The TIN shown in Figure 12.2 apparently seems to contain far more short edges than long ones. This is indeed true. There are 504 edges, ranging from the shortest 0.001 to the longest 46.752. The wide range, 46.751 = 46.752 –0.001, and the large ratio, 46,752 = 46.752/0.001, clearly indicate far more short edges than long ones, although the exact distribution is up to further statistic investigation. The average length of the 504 edges is 2.2, which splits all the edges into two unbalanced parts: 135 in the head (27%) and 369 in the tail (73%). This head/tail-breaks process continues for the head again and again, as shown in Table 12.1. Eventually, the scaling pattern of far more short edges than long ones recurs five times, three of which are plotted in Figure 12.3, or so-called nested rank-size plots. Given that the scaling pattern recurs five times, the head/tail breaks, or ht-index, is six. Note that the ht-index (Jiang & Yin, 2014) is an alternative index to fractal dimension (Mandelbrot, 1983) used to capture the complexity of geographical features.

Table 12.1 Head/tail breaks statistics for the TIN edges

# Edges	Mean	# Head	% Head	# Tail	% Tail
504	2.2	135	27%	369	73%
135	6.2	35	26%	100	74%
35	13.4	13	37%	22	63%
13	20.7	3	23%	10	77%
3	33.2	1	33%	2	67%

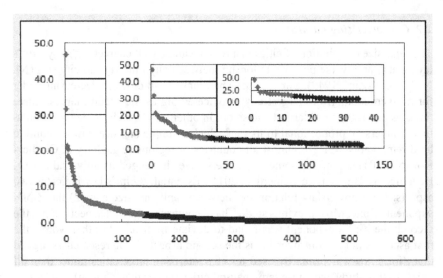

Figure 12.3 (Colour online) Nested rank-size plots for the first three hierarchical levels with respect to the first three rows in Table 12.1

Note: The x-axis and y-axis represent rank and size, respectively. The largest plot contains the 504 edges, the red being the first head (135 edges) and the blue being the first tail (369 edges). The 135 edges are plotted again with the red representing 35 in the second head and the blue 100 in the second tail. The smallest plot is for the 35 edges in the second head.

The ht-index provides a simple yet effective means for characterizing natural cities or data, in general, with a heavy-tailed distribution for mapping purposes. The derived ht-index captures well the hierarchy or scaling hierarchy of the data. For mapping purposes, the head/tail breaks is superior to conventional classification methods for capturing the underlying scaling pattern (Jiang, 2013, 2015a). Ht-index complements to fractal dimension for characterizing the complexity of geographic features or fractals in general.

3. Data and data processing

As stated earlier, the data for this study came from the former location-based social medium Brightkite, during its three-year (31 months to be more precise) life span, from April 2008 to October 2010 (Cho, Myer, & Leskovec, 2011). The case included 2,788,042 locations in the contiguous United States. From the number of locations, we removed duplicate locations, obtained 404,174 unique locations for generating a TIN and then 8,106 natural cities as of October 2010, by following the procedure shown in Figure 12.2, as well as the short tutorial presented in the Appendix at the end of this chapter. The location data was timestamped (Table 12.2), so we were able to slice all these locations monthly in an accumulated manner, that is locations at month m_{i+1} contain all locations between months m1 and mi, where $1 \leq i \leq 31$. For each time interval or snapshot, we generated a set of natural cities ranging from 100 to 8,106. Figure 12.4 illustrates the 8,106 natural cities as of October 2010, showing their boundaries and populations. Note that this is just one of the 31 snapshots or datasets in the study.

Table 12.3 lists some basic measurements and statistics from the location points of the natural cities. For example, for the first month, April 2008, only 100 natural cities were generated from 8,563 locations, of which 3,007 unique locations were used for generating a TIN with 9,001 edges, and a mean of 33,775 as the cut-off to derive the 100 natural cities. The number of natural cities increased steadily to 8,106 as of October 2010. In the following section, we utilize the seven time intervals highlighted in Table 12.3 for a detailed discussion of our findings.

4. Results and discussion

Before discussing the results, we map the natural cities at the seven time intervals (or snapshots) for the four largest natural cities surrounding Chicago, New York,

Table 12.2 Initial check-in data format

User	Check-in time	Latitude	Longitude	Location id
58186	2008–12–03T21:09:14Z	39.633321	−105.317215	ee8b88dea22411
58186	2008–11–30T22:30:12Z	39.633321	−105.317215	ee8b88dea22411
58186	2008–11–28T17:55:04Z	−13.158333	−72.531389	e6e86be2a22411
58186	2008–11–26T17:08:25Z	39.633321	−105.317215	ee8b88dea22411
58187	2008–08–14T21:23:55Z	41.257924	−95.938081	4c2af967eb5df8

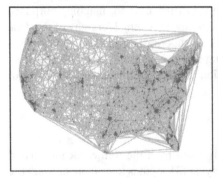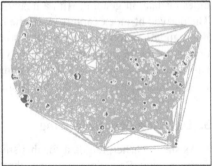

Figure 12.4 (Colour online) The largest set of natural cities as of October 2010 (red patches for boundaries and red dots for populations) on the background of TIN (grey lines) generated from 404,174 unique location points or 2,788,042 duplicate ones

Table 12.3 Measurements and statistics from location points to natural cities for the different time intervals

Time	Pnt	PntUniq	TINEdge	Mean	NaturalCity
2008–04	**8,563**	**3,007**	**9,001**	**33,775**	**100**
2008–05	**104,848**	**21,171**	**63,492**	**10,589**	**437**
2008–06	207,048	35,176	105,505	7,910	698
2008–07	308,987	46,447	139,317	6,882	938
2008–08	**405,866**	**57,458**	**172,351**	**6,129**	**1,147**
. . .					
2008–12	**903,990**	**135,998**	**407,970**	**3,636**	**2,558**
. . .					
2009–06	**1,800,825**	**263,103**	**789,284**	**2,518**	**5,132**
. . .					
2009–12	**2,407,118**	**361,990**	**1,085,945**	**2,109**	**7,243**
. . .					
2010–10	**2,788,042**	**404,174**	**1,212,498**	**1,969**	**8,106**

Note: Pnt = # of points, PntUniq = # of unique points, TINEdge = # of TIN edges, Mean = Average length of TIN edges, NaturalCity = # of natural cities. The seven rows highlighted in bold represent the seven snapshots in Figures 12.5, 12.6 and 12.7.

San Francisco and Los Angeles. These are shown in Figure 12.5, which illustrates clearly how the four regions grew during the 31-month period. All parts of the country can be assessed for similar patterns of growth and evolution. We know little why the procedure illustrated in Figure 12.2 and the Appendix, works so well, but the resulting patterns suggest that the natural cities effectively capture the evolution of real cities. On the one hand, the natural cities expanded towards more fragmented pieces, far more small pieces than large ones. On the other hand, the physical boundaries of the natural cities tended to become more irregular and finer

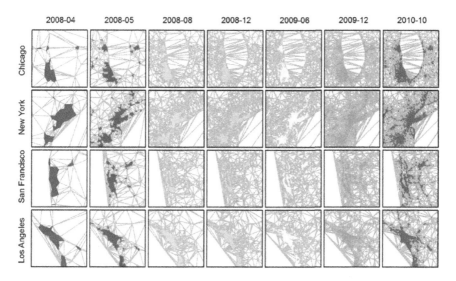

Figure 12.5 (Colour online) Evolution of the natural cities near the four largest cities regions with TIN as a background

over time. These two aspects suggest that the natural cities are fractal and become more and more fractal, resembling real cities based on conventional population data (Batty & Longley, 1994).

These results can be assessed from both global and local perspectives. Globally, all the natural cities in the contiguous United States exhibit a power-law distribution. This is shown rank-size plots (Figure 12.6), in which the distribution lines are very straight for all the natural cities at different time intervals in the log-log plots. The natural cites, as of April 2008, except the smallest with fewer than 12 people, exhibited a clear power law, probably the straightest distribution among all others. However, the distribution lines from May 2008 to October 2010 are less straight, indicating that a few of the largest natural cities did not fit the power-law distribution very well. This is particularly obvious for the last two snapshots in December 2009 and October 2010. A possible reason for this difference, moving from a striking to a less striking power law, is described in the following.

In further examinations, we looked at the large cities (larger than the mean) in each snapshot and found that Zipf's exponent was indeed around one for the first month (0.99), and then greater than one by about 0.3 (Table 12.4). Considering the duality of Zipf's law, this result suggests that Zipf's law held remarkably well for the first month, but less so for the remaining months. We postulated a possible reason: The social medium users at the first month increased proportionally with the populations of real cities, thus leading to a striking Zipf's law effect among the natural cities because the populations of real cities are power-law distributed. Over time, large cities – particularly a few of the largest cities, such as New York – did

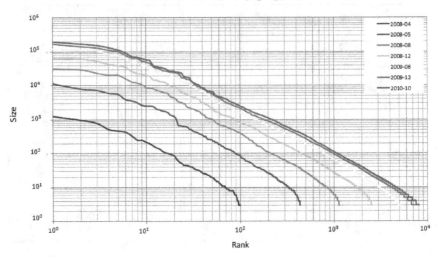

Figure 12.6 (Colour online) Rank-size plot for the natural cities

Table 12.4 Zipf's exponent and ht-index for the natural cities

	2008–04	*2008–05*	*2008–08*	*2008–12*	*2009–06*	*2009–12*	*2010–10*
Zipf's exponent	0.99	1.19	1.25	1.28	1.29	1.28	1.26
Ht-index	2	4	4	4	4	5	5

not capture the other cities in attracting more users. In other words, beyond the first month, the increase in social media users became less proportional to the real cities' populations. As a result, Zipf's law is less striking. In contrast to small deviations of Zipf's exponent, the ht-index increased from 2 to 5 (Table 12.4). The increment of the ht-index implies that more hierarchical levels were added, reflecting well the evolution of the natural cities and of the social medium.

Locally, there are two points to discuss. First, the boundaries of the natural cities became more irregular and finer over time, very much like the Koch curve when the iteration goes up. For example, the boundaries of the natural cities as of April 2008 were simple enough to be described by Euclidean geometry. However, over time, the boundaries must be characterized by fractal geometry – more fragmented with more fine scales added. Second, large natural cities tended to become larger and larger, while small ones continuously emerged at local levels. Figure 12.5 illustrates this finding in a less striking manner, as the city sizes are measured by the physical extents. But if the city sizes are measured by the users (or population) as in Figure 12.7, we noticed the rapid increases for the four largest cities. Overall, the four cities tended to become larger and larger, but there was

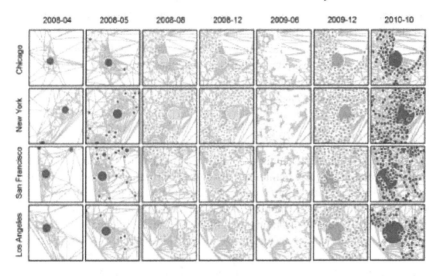

Figure 12.7 (Colour online) Evolution of the natural cities in terms of populations (or points) near the four largest cities regions

Table 12.5 Evolution of the four cities within the system of the natural cities

	2008–04	2008–05	2008–08	2008–12	2009–06	2009–12	2010–10
Chicago	3/4	4/5	4/5	5/6	5/6	5/6	6/7
New York	3/4	5/5	5/5	5/6	6/6	6/6	7/7
San Francisco	4/4	5/5	5/5	6/6	6/6	6/6	7/7
Los Angeles	4/4	5/5	5/5	6/6	6/6	6/6	7/7

Note: a/b, where a and b, respectively, denote the class the particular city belongs to, and the total number of classes or the ht-index.

a major difference among the four. To illustrate the difference, we must clarify that Figure 12.7 adopts the graduated dots to represent the city sizes, which are classified according to head/tail breaks. This is because the city sizes exhibited a heavy-tailed distribution, or there were far more small cities than large ones. Therefore, the dot sizes in Figure 12.7 do not represent city sizes, strictly speaking, but, rather, the corresponding classes to which the cities belong. Notice that the largest natural city in the Chicago region in October 2010 appears smaller than others. This is indeed true! Table 12.5 clearly indicates that the Chicago natural city in October 2010 belongs to the sixth among the seven classes, while the others belong to the seventh among the seven classes.

The preceding results can be summarized by one word – non-linearity, which is reflected in both spatial and temporal dimensions. Spatially, the natural cities were distributed heterogeneously or unevenly; that is there were far more small cities than large ones. This uneven distribution also was seen in the temporal dimension. For example, within the first nine months of 2008, the natural cities had already been shaped (Figure 12.5), with the users or populations continuously growing and small natural cities being added persistently for the remaining time. In other words, it took just one third of the social medium's lifetime to determine the shapes of individual cities. That is also the reason that we chose the seven unequal time intervals to examine the evolution.

5. Implications of the study

The location-based social media provide large amounts of location data of significant value for studying human activities in the virtual world, as well as on the Earth's surface. Nowadays, the social sciences – human geography, in particular – benefit considerably from emerging social media data that are timestamped and location-based. The ways of doing geography and social sciences are changing! The emerging big data harvested from social media, as well as from geospatial technologies, coupled with data-intensive computing (Hey, Tansley, & Tolle, 2009) are transforming conventional social sciences into computational social sciences (Lazer et al., 2009). In this section, we discuss some deep implications of this study for geography and social sciences in general.

The notion of natural cities implies a sort of bottom-up thinking in terms of data collection and geographic units or boundaries. Conventional geographic data collected and maintained from the top down by authorities are usually sampled and aggregated, and therefore, are small-sized. On the other hand, new data harvested from social media are massive and individual, so they are called "big data". Timestamped and location-based social media data, supported by Web 2.0 technologies and contributed by individuals through humans as sensors (Goodchild, 2007), constitute a brilliant new data source for geographic research. Conventional geographic units or boundaries are often imposed from the top down by authorities or centralized committees, while natural cities are defined and delineated objectively in some natural manner, based on the head/tail breaks. This natural manner guarantees that we can see a true picture of urban structure and dynamics and suggests the universality of Zipf's law. This true picture is fractal and can be illustrated in this example: forcefully throw a wine glass on a cement ground, and it will very likely break into a large number of pieces. Like natural cities, these glass pieces are fractal or follow Zipf's law: on one hand, there are far more small pieces than large ones, and on the other, each piece has an irregular shape.

The evolution of natural cities demonstrates non-linearity at both spatial and temporal dimensions, or equivalently from both static and dynamic points of view. Many phenomena in human geography, as well as in physical geography, bear this non-linearity (Batty & Longley, 1994; Frankhauser, 1994; Chen, 2009; Phillips, 2003). However, we are still very much constrained by linear thinking, explicitly or implicitly, consciously or unconsciously. For example, we rely on Euclidean geometry to describe Earth's surface, and on a well-defined mean to characterize spatial heterogeneity. Our mind-sets apparently lag the advances of data and technologies. Conventional linear thinking is not suitable for describing the Earth's surface (the geographic forms), not to mention uncovering the underlying geographic processes. Instead, we should adopt non-linear thinking or non-linear mathematics, such as fractal geometry, chaos theories and complexity for geographic research. The tools adopted in this study, such as head/tail breaks and ht-index, underlie non-linear mathematics and power-law-based statistics. These non-linear mathematical tools help to elicit new insights into the evolution of natural cities. Non-linearity also implies that geographic forms and processes are unpredictable like long-term weather or climate, in general. To better predict and understand geographical phenomena, we must seek to uncover the underlying mechanisms through simulations rather than simple correlations.

The head/tail breaks is intellectually exciting because it appears to be both powerful and mysterious. The reason why the head/tail breaks is an effective tool to derive natural cities, in particular at the different time stages, remains an open question. However, we tend to believe it is the effect of the wisdom of crowds – the diverse and heterogeneous many are often smarter than the few, even a few experts (Surowiecki, 2004). The massive amount of edges (up to 1,212,498) of the generated TIN from the massive location points constituted the "crowds", and they collectively decided an average cut-off for delineating the natural cities. Every single edge had "its voice heard" in the democratic decision. From the effectively derived natural cities, we can see an advantage of working with big data. If we had not worked with the entire U.S. data set, but only an area surrounding New York for example, we would not have been able to determine a sensible cut-off for delineating the New York natural city. Only with the big data that include all location points or all edges can a meaningful cut-off be determined and applied to all. In this sense, the approach to delineating natural cities is holistic and bottom-up, with the participation of all diverse and heterogeneous individuals.

It is important to note that check-in users are biased towards certain types of people. Thus, derived natural cities are not exactly the same as the corresponding real cities. However, no one can deny that the boundaries shown in Figure 12.5 are not those of Chicago, New York, San Francisco and Los Angeles. Note that this chapter is not to study real cities, at an individual level, regarding how they can be captured or predicted by natural cities but to understand, at a collective level, underlying mechanisms of agglomerations, formed either by people in physical space (real cities) or by the check-in users in virtual space (natural cities). In other words, we consider cities (either real or natural cities) as an emergence (Johnson, 2002) developed from the interaction of individual people or that of cities. The fractal structure and non-linear dynamics illustrated appear to be applied to both real and natural cities. The fact that not all people are the check-in users should not be considered a biased sampling issue. Sampling is an inevitable technique at the time of information scarcity, the so-called small data era, but it is not a legitimate concept in the big data era. The large social media data imply $N = $ all (Mayer-Schonberger & Cukier, 2013). This $N = $ all is an essence of big data. Given the 2.8 million of check-in locations, social media can be a good proxy for studying the evolution of real cities in the country.

We face an unprecedented golden era for geography, or social sciences in general, with the wave of social media and, in particular, the increasing convergence of social media and geographic information science (Sui & Goodchild, 2011). For the first time in history, human activities can be documented at very fine spatial and temporal scales. In this study, we sliced the data monthly, but we certainly could have done so weekly, daily and even hourly. We believe that the observed non-linearity at the temporal dimension would be even more striking. This, of course, warrants further study. Geographers should ride the wave of social media and develop a more computationally minded geography or computational geography (Openshaw, 1998). If we do not seize this unique opportunity, we may risk being purged from the sciences. The rise of computational social science is a timely response to the rapid advances of data and technologies. In fact, physicists and computer scientists are already working in this exciting and rapidly changing domain (see Brockmann, Hufnage, & Geisel, 2006; Zheng & Zhou, 2011). We geographers should do more rather than less.

6. Conclusion

Driven by the lack of data for tracking the evolution of cites, this study demonstrated that emerging location-based social media such as Flickr, Twitter, and Foursquare can act a proxy for studying and understanding underlying evolving mechanisms of cities. Compared with conventional census data that are usually sampled, aggregated, and small, the timestamped and location-based social media data can be characterized as all, individual, and big. In this chapter, we abandoned conventional definitions of cities; adopted objectively or naturally defined natural cities, using massive geographic information of various kinds; and based on the head/tail breaks. Built on the notion of the wisdom of crowds, the head/tail breaks works very well to establish a meaningful cut-off for delineating natural cities. Natural cities provide an effective means or unique perspective to study human activities for a better understanding of geographic forms and processes.

We examined the evolution of natural cities, derived from massive location points of the social media platform Brightkite, during its 31-month life span. We found non-linearity during the evolution of natural cities in both spatial and temporal dimensions, and the universality of Zipf's law. We archived all the data that could be of further use for developing and verifying urban theories. This study has deep implications for geography and social sciences in light of the increasing amounts of data that can be harvested from location-based social media. Therefore, we call for the application of non-linear mathematics, such as fractal geometry, chaos theories and complexity to geographic and social science research. A limitation of this study lies in the data that show only social media's continuous rise and not its decline. Brightkite seemed to disappear overnight. Future research should concentrate on the development of power-law-based statistics and the underlying non-linear mathematics to manage the increasing social media data and on agent-based simulations to reveal the mechanisms for the evolution of natural cities.

Acknowledgement

This chapter is a reprint of the journal article (Jiang and Miao 2015) with the permission kindly granted by the publisher Taylor & Francis Ltd. An early version of this chapter was presented by the first author as a keynote titled "The evolution of natural cities: a new way of looking at human mobility", at Mobile Ghent '13, 23–25 October 2012, University of Ghent, Belgium. We would like to thank Kuan-Yu Huang for partial data processing and the three anonymous referees and editor for their comments that significantly improved the quality of this chapter.

References

Bak, P. (1996). *How nature works: The science of self-organized criticality.* New York, NY: Springer-Verlag.

Batty, M., & Longley, P. (1994). *Fractal cities: A geometry of form and function.* London: Academic Press.

Benguigui, L., & Blumenfeld-Leiberthal, E. (2011). The end of a paradigm: Is Zipf's law universal? *Journal of Geographical Systems, 13,* 87–100.

Bennett, J. (2010). *OpenStreetMap: Be your own cartographer.* Birmingham: PCKT Publishing.

Berry, B. J. L., & Okulicz-Kozaryn, A. (2011). The city size distribution debate: Resolution for US urban regions and megalopolitan areas. *Cities, 29,* S17-S23.

Boyd, D. M., & Ellison, N. B. (2008). Social network sites: Definition, history, and scholarship. *Journal of Computer-Mediated Communication, 13,* 210–230.

Brockmann, D., Hufnage, L., & Geisel, T. (2006). The scaling laws of human travel. *Nature, 439,* 462–465.

Chen, Y. (2009). Spatial interaction creates period-doubling bifurcation and chaos of urbanization. *Chaos, Solitons & Fractals, 42*(3), 1316–1325.

Cho, E., Myers, S. A., & Leskovec, J. (2011). Friendship and mobility: User movement in location-based social networks. *Proceedings of the 17th ACM SIGKDD International Conference on Knowledge Discovery and Data Mining* (pp. 1082–1090). ACM. New York.

Frankhauser, P. (1994). *La Fractalité des Structures Urbaines.* Paris: Economica.

Goodchild, M. F. (2007). Citizens as sensors: The world of volunteered geography. *Geo-Journal, 69*(4), 211–221.

Hey, T., Tansley, S., & Tolle, K. (2009). *The fourth paradigm: Data intensive scientific discovery.* Redmond, WA: Microsoft Research.

Jia, T., & Jiang, B. (2010). Measuring urban sprawl based on massive street nodes and the novel concept of natural cities. Preprint: http://arxiv.org/abs/1010.0541 (Accessed October 7, 2014).

Jiang, B. (2013). Head/tail breaks: A new classification scheme for data with a heavy-tailed distribution. *The Professional Geographer, 65*(3), 482–494.

Jiang, B. (2015a). Head/tail breaks for visualization of city structure and dynamics, *Cities, 43,* 69–77. Reprinted in Capineri, C., Haklay, M., Huang, H., Antoniou, V., Kettunen, J., Ostermann, F., & Purves, R. (Eds.). (2016). *European handbook of crowdsourced geographic information* (pp. 169–183). London: Ubiquity Press.

Jiang, B. (2015b). Geospatial analysis requires a different way of thinking: The problem of spatial heterogeneity. *GeoJournal, 80*(1), 1–13. Reprinted in Behnisch, M., & Meinel,

G. (Eds.). (2017). *Trends in spatial analysis and modelling: Decision-support and planning strategies* (pp. 23–40). Berlin: Springer.

Jiang, B., & Jia, T. (2011). Zipf's law for all the natural cities in the United States: A geospatial perspective. *International Journal of Geographical Information Science, 25*(8), 1269–1281.

Jiang, B., & Liu, X. (2012). Scaling of geographic space from the perspective of city and field blocks and using volunteered geographic information. *International Journal of Geographical Information Science, 26*(2), 215–229. Reprinted in Akerkar, R. (Ed.). (2013). *Big data computing* (pp. 483–500). London: Taylor & Francis.

Jiang, B., & Miao, Y. (2015). The evolution of natural cities from the perspective of location-based social media. *The Professional Geographer, 67*(2), 295–306.

Jiang, B., & Yin, J. (2014). Ht-index for quantifying the fractal or scaling structure of geographic features. *Annals of the Association of American Geographers, 104*(3), 530–541.

Johnson, S. (2002). *Emergence: The connected lives of ants, brains, cities, and software.* New York, NY: Scribner.

Kaplan, A. M., & Haenlein, M. (2010). Users of the world, unite! The challenges and opportunities of social media. *Business Horizons, 53*, 59–68.

Lazer, D., Pentland, A., Adamic, L., Aral, S., Barabási, A.-L., Brewer, D. . . . Van Alstyne, M. (2009). Computation social science. *Science, 323*, 721–724.

Mandelbrot, B. (1983). *The fractal geometry of nature.* New York, NY: W. H. Freeman and Co.

Mayer-Schonberger, V., & Cukier, K. (2013). *Big data: A revolution that will transform how we live, work, and think.* New York, NY: Eamon Dolan/Houghton Mifflin Harcourt.

Openshaw, S. (1984). *The modifiable areal unit problem.* Norwick, Norfolk: Geo Books.

Openshaw, S. (1998). Towards a more computationally minded scientific human geography. *Environment and Planning A, 30*, 317–332.

Phillips, J. D. (2003). Sources of nonlinearity and complexity in geomorphic systems. *Progress in Physical Geography, 27*(1), 1–23.

Sui, D., & Goodchild, M. (2011). The convergence of GIS and social media: Challenges for GIScience. *International Journal of Geographical Information Science, 25*(11), 1737–1748.

Surowiecki, J. (2004). *The wisdom of crowds: Why the many are smarter than the few.* London: Abacus.

Traynor, D., & Curran, K. (2012). Location-based social networks. In I. Lee (Ed.), *Mobile services industries, technologies, and applications in the global economy* (pp. 243–253). Hershey, PA: IGI Global.

Zheng, Y., & Zhou, X. (Eds.). (2011). *Computing with spatial trajectories.* Berlin: Springer.

Zipf, G. K. (1949). *Human behavior and the principles of least effort.* Cambridge, MA: Addison Wesley.

Appendix

Tutorial on how to derive natural cities based on ArcGIS

This tutorial aims to show, in a step-by-step fashion with ArcGIS, how to derive natural cities using the first month data (2008–04) as an example. Once you have got the check-in data of the first month, transfer them into an Excel sheet with two columns, namely x, y, respectively representing longitude and latitude. Add a third column, z, and set all column values as one (or any arbitrary value since ArcGIS relies on three-dimensional points for creating a TIN). Insert the Excel sheet as a shape file data layer in ArcGIS (Figure 12.A1(a)). Create a TIN from the point layer, using ArcToolbox > 3D analyst tools > TIN management > Create TIN (Figure 12.A1(b)).

Convert the TIN into TIN edge, using ArcToolbox > 3D analyst tools > Conversion > From TIN > TIN Edge. The converted TIN edge is a polyline layer. Right

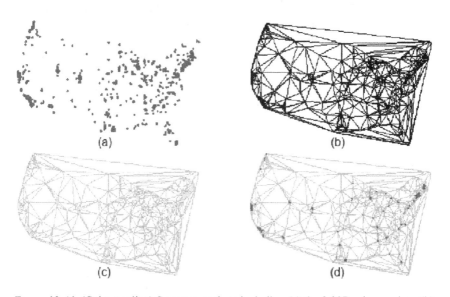

Figure 12.A1 (Colour online) Screen snapshots including (a) the 3,007 unique points, (b) the TIN from the 3,007 points, (c) the selected edges shorter than the mean 33,775 and (d) the 100 natural cities created

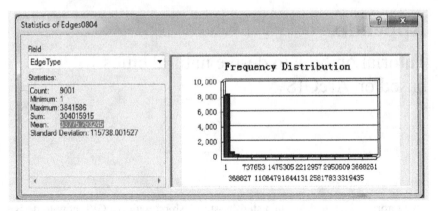

Figure 12.A2 (Colour online) Statistics of TIN edges

click the polyline layer to Open Attribute Table in order to get statistics about the length of the edges. Figure 12.A2 shows that the frequency distribution that is apparently L-shaped, indicating that there are far more short edges than long ones. Note that the mean is 33,775.

Select those shorter edges than this mean 33,775 following menu Selection > Select by Attributes. . . . The selected edges are highlighted in Figure 12.A1(c). The selected shorter edges refer to high-density locations. Dissolve all the shorter edges into polygons to be individual natural cities (Figure 12.A1(d)), following ArcToolbox > Data Management Tools > Features > Feature to Polygon > Generalization > Dissolve, or alternatively following menu Geoprocessing > Dissolve (the option "Create multipart features" should be unchecked).

The preceding steps all can be done with the existing ArcGIS functions. The reader may encounter some problem with more check-in points are added cumulatively. In this case, you must spit the data into small pieces and put them back again into ArcGIS with some simple codes. All the produced data are archived at https://sites.google.com/site/naturalcitiesdata/.

Epilogue, looking into the near future of information and communication technology–enabled travel

Anne Vernez Moudon

> *"Physiologically, man in the normal use of technology (or his variously extended body) is perpetually modified by it and in turn finds ever new ways of modifying his technology." When new technologies impose themselves on societies long habituated to. . . . Man becomes, as it were, the sex organ of the machine world, as the bee of the plant world, enabling it to fecundate and to evolve ever new forms." "The machine world reciprocates man's love by expediting his wishes and desires, namely, in providing him with wealth."*
>
> —M. McLuhan, *Understanding Media* (1964, p. 46)

Information and Communication Technologies (ICT) has speared a revolution in travel that is hard to grapple with. More precisely, however, it is *mobile* Information and Communication Technology (m-ICT) that is radically altering travel (Campbell, 2013). Today's mobile technology, characterized by application (app)–enabled smartphones, irrevocably links travel to social network and institutional structures. In economic terms, it connects travel demand with supply options in a "real" space-time continuum.

The boundaries among ICT, m-ICT, social media and social networks are increasingly blurred. Yet, as McLuhan described it long ago (McLuhan, 1964), technology is the force that is pulling society into the next generation of travel behaviours and habits. One would wish for a reverse situation, one where society was using technology to direct it into new ways of travel or where society was "pushing" in the desired direction rather than being "pulled" in a more random direction. This wish is still unfulfilled. Taming the technology, and having it serve us rather having us serving it, is at the centre of this book's contributions.

Mobile ICT in transport

At a general level, parcelling out the effects of mobile ICT on travel shows clear trends. On the demand side, smartphones have provided individual owners with far better control of their mobility than they had in the past: travellers, whether they be solo drivers or transit users, have ready access to information on route and travel mode, including (importantly) access to information about price and real time of service. Indeed, with an investment in a smartphone and a forfeit to a service provider, an individual can benefit from *temporally and spatially continuous* access to travel information services at relatively low cost. These services now act as a virtual extension of the physical streets and roads that cities offer – also at low cost – to house our travels. Smartphone users can locate the nearby transit stops, parking garages, and so forth that their eyes may yet be able to see, and they can anticipate with relative precision when the bus, train or private ride they plan to take will reach them. On the supply side, m-ICT has extended the range of the shared economy to include novel options in transport. Transportation network companies (TNCs) or mobility service providers (MSPs) such as Uber, Car2go, and other local companies, have emerged, and their ability to communicate directly with users has radically changed the choices travellers can make. In cities especially, shared vehicles (cars, bicycles, scooters, etc.), together with ride-hailing services, offer promises of increased mobility at seemingly affordable prices. Phone-based access to shared vehicles provides new flexibility in mode choice at the day level – a person can mix transit and single-vehicle travel for the work commute, he or she can opt for walking or bicycling depending on weather and so on. Flexibility is the most tangible because the need for private vehicle storage (aka parking) is removed, along with the burden of having to secure or pay for such storage.

With m-ICT, travellers are connected, and indeed interconnected, with a range of interested parties in public agencies (e.g. Departments of Transport) or private services (including TNCs), as well as with ICT and related location-based communication service companies. Most, if not all, participants in the vast travel information networks have a presence on social media, and many communications take place via social media platforms in addition to individual apps.

Travel information: the public/private data divide

Armed with smartphones, travellers are at the centre of an important information exchange. As they pay for their physical phone, for a connection to a network provider with a data plan, travellers may not fully appreciate that the part of what is still called a "phone bill" not only serves to facilitate their mobility options. It also generates records of their mobility patterns, which remain in the control of TNCs and location-based ICT companies. As TNCs typically do not share their mobility data, public transport agencies managing transport infrastructure are left to react to the effects of new mobility options (e.g. ride hailing, bike share) on individual car and transit travel: Are they substituting for or complementing transit trips? Are they catering to solo travellers or to groups of travellers? Similarly, private location-based ICT companies monitoring vehicle-based mobility (e.g. INRIX, Waze) sell their data to public agencies. Yet the data have severe limitations because they come with minimal descriptions of the methods used for data collection and analysis, making it difficult to use them in statistical models guiding public policy and to make cogent inferences from the results of analyses.

More broadly, to understand the new shared mobility trends, to monitor changes, and eventually to properly plan transport systems, it will be essential to harness the wealth of data available on travellers' smartphones for everyone's and anyone's to share. We are in a situation that seems to be similar to that of the late 19th and early 20th centuries: many cities became overwhelmed with disjointed metro and streetcar systems that had been privately developed through concessions. At that time, inefficiencies and conflicts between private transport providers, whose physical networks had been installed primarily to access developable land away from city centres, had to be eventually taken over by the public sector after the First World War. In light of this history, one can argue that today's TNCs and location-based services are using transport not to reach out to undeveloped lands but to collect data on people's mobility patterns in order to further their grip on advertising and consumer spending – an issue addressed again later in this chapter. However, TNCs, like private transport companies in the early 20th century, are realizing that transport involves many participants. Those in the ride-hailing and car- or bike-share business directly bear the cost of inefficiencies in transport systems. Traffic congestion and air pollution propagated by the predominance of individual car travel or by the inadequacy of the transit system also have a negative impact on shared mobility options because they too are "stuck in traffic". TNCs are acutely aware that mass mobility cannot be served by one transport mode and that, in cities, they could only benefit from well-functioning road- or rail-based collective modes. As a group, TNCs are realizing that in order to fully benefit all mobility, their data should be extended to others and properly shared. As individual companies, however, TNCs are unlikely to take the first step and share data with others. Their narrow competitive streak leads them to strive to dominate the market and not to collaborate with others. It seems that policy- and regulation-making bodies who control the transport infrastructure and the utilities may be best positioned to play a role in reining in and requiring data sharing for all users of the public infrastructure. This leads to posing a more reasonable question: how long will vehicular traffic congestion continue before governments and regulators seek to tame the ICT industries involved in surface transport? We will return to this question after reflecting on what kinds of entities the TNCs and location-based services are.

Transport as a means but not an end for global ICT: big data and the vertical integration of transport

TNCs are profit-seeking private companies. They are misnamed because transport is only a means to their actual end, the acquisition of mobility data. Data-hungry ICTs emerged in the late 1990s as the dot-com bubble propelled surviving information technology (IT) companies into coming up with new, cash-generating business models. After 11 September 2001, some of them were invited by governments to step into the area of security and surveillance. The dramatic events marked the beginning of the now irreversible link between IT and transport. Cybersecurity and air transport security came as first concerns, only to be followed rather quickly by a focus on surface transport for both people and goods. Long before then, banks, merchants, social media and others had already monetized the behaviour information they were gathering via credit cards, loyalty cards and websites, and had used them to sell advertisement, ideas and products. New "security" concerns began to demand information on people's mobility – information regarding their location, activity and social behaviour. Advances in mobile and smartphone-based technology and its accelerated use only exacerbated the rapprochement between the ICT and transport sectors. Travellers could be connected to cell towers and located in real time, beaming continuous strings of data back to the ICT industry. These were part of the big data that has since fascinated every sector, including transport. The data have been euphemistically labelled "social data" and defined as information that social media users *publicly share* – italics added.

Many m-ICT-enabled shared transport providers are global companies whose financiers operate at the world market level. Uber left the Chinese market to the Chinese company Didi but remains a sizeable Didi shareholder. At the time of this writing, Uber Technologies Inc. has officially filed for an initial public offering with the New York Securities and Exchange Commission. Overall, global mobility service providers are remarkably well-capitalized – especially as compared to perennially underfunded local public transport entities. With relatively low extra investment, some have moved to acquire bike-share businesses which, if successful, would let them capture a large share of local mobility patterns from home, work and other places. Vertically integrated global private transport providers will also be able to control the cost of transport (The Economist, 2018). It seems evident that pending unexpected changes in the economy, global ICTs are in the driver's seat (pun intended) of future transport worldwide.

The value of travel: travellers as enablers of "free" tech

The monetary value of the big data handled by mobility and location-based service providers has yet to be fully appreciated by the transport sector, let alone precisely estimated. Travellers themselves do not realize that their monetary contribution to their travel, in fact, pales in comparison with the value of the information that their smartphones transmit to the "world" regarding their location, activity and social behaviour. According to the theory of *surveillance capitalism* (Zuboff, 2015, 2019), spatio-behavioural data gleaned from smartphones have become an essential input to the prediction of future human behaviour. Specifically, big data form a closed behaviour-information loop where, as explained long ago by classic social and behavioural science theory, the observation of behaviour serves as basis for prediction, which then provides information on the determinants of behaviour, which, in turn, can be used to identify the mechanisms for behaviour change, leading eventually to *expected modified behaviour*. Critics of surveillance capitalism accuse the large ICT companies to practising the theory, first, to derive value from their users' data and, second, to manipulate their users' behaviour for even bigger gains. "Surveillance capitalism's real products, vaporous but immensely valuable, are predictions about our future behaviour . . . these new data derivatives draw their value, parasite-like, from human experience" (Carr, 2019). As an aside, the fine-line difference between scientific research to inform behaviour and promote behaviour change and ICT manipulation of behaviour for self-gain lies in the expression of intent, full disclosure and accountability on the part of researchers but not industry and of the subjects' consent that scientists must obtain from any and all subject of research (the result of many years of struggle to define an ethics of studying the "other").

Whatever side of surveillance capitalism one might be, the point is that smartphone-assisted travellers are not mere users of the "free" technology; they also are its enablers. At this time, the traveller's behaviour is a valuable "data product" used mostly by private mobility providers. Governments and the transport sector need to intervene, to extract value from the m-ICT companies involved in how people travel, to acquire access to the data, and to apply them comprehensively to the entire transport system. Making the travelling public more aware of the value of the data they generate will be a first step in convincing TNCs and location-based companies of the need to openly share the data.

Government versus global ICT companies

As urban history has shown, efficient transport must be treated as a shared if not a public resource, where the public sector is more than the investor and builder of the transport infrastructure, it is also its *de facto* guardian. The rise of global ICT companies in the transport sector raises questions as to how well-equipped governments are to negotiate with large corporate entities. Local governments typically consider global companies as employers, which they need to court for economic development. As TNCs increasingly come from the global corporate world, local politicians will negotiate with them on labour issues (e.g. Uber drivers vs taxi drivers) before they are ready to do so on issues directly related to transport. Only the richest and most forward-looking cities are able to impose their will on private transport providers. Few have fought the car industry's claim on urban public space – and the few that have been successful at reducing the impact of individual vehicles in their midst typically only did so after environmental and congestion issues had become dire. As exemplified by the EU General Data Protection Regulations, reining into global corporations on tax and privacy issues is best achieved at higher levels of government. However, even at these higher levels of government, the most effective lever used to achieve success in the transport sector has been through concerns for the environment, and notably through addressing the impacts of ill-advised transport policies and options on air quality. Looking into the future, the broader lens of health promotion – encouraging active transport, noise-free and clean air environment – might become useful in guiding forthcoming transport policies and investments.

Day-to-day travel with m-ICT: issues and potentialities

A series of questions remain on how well m-ICT is serving the travelling public. Some of the following suggest the need for further research and development.

Costs to the user

Being connected to the web has become an added direct cost to individual transport. Travellers now need to invest in a smartphone and subscribe to related services. The European Union has tried to contain these costs, but in other parts of the world, the substantial costs of ICT-enabled travel compete with the end user's cost of housing, food and education. There are indirect costs as well. Smartphones (as most computers) operate by self-help, which, on one hand, provides the user with a sense of control yet, on the other hand, consumes large amounts of time that will be added to the travel "budget." For example, travellers must choose between multiple websites for train or plane tickets – only the rich have access to travel agent services. The multiple sites will exist as long as individual businesses benefit from customers using their own apps. There is a need for apps that serve the user rather than specific companies or agencies.

Travel is more than a single origin–destination (O/D) trip

Is there an online route type of app catering to individuals travelling in urban areas? Google Maps are ahead of others for individual route optimization services. They provide an increasingly large number of options by mode (i.e. metro, bus, Uber, etc.) and let users control the route taken between a trip origin and destination. Their options are, first, offered by travel duration and primary mode (i.e. individual vehicle, transit, active travel); access modes to transit (typically walking) are also shown, providing important information. The app recently started to offer the possibility of incrementally adding stops to a trip; stops appear on the map, along with directions and total travel duration based on the selected primary travel mode. The next step is likely for Google Maps to provide the ability to plan a multimodal itinerary with multiple stops. Transit-specific apps do not accommodate chain-linked trips. Only air travel can be planned with multiple destinations and multiple carriers. Intercity rail travel can also include multiple stops. However, long-distance travel apps typically consider single primary modes (e.g. plane or train) and only provide the choice of a few options (e.g. bus or active travel) for access to primary modes. Of note, prices are often higher on multicity trips, suggesting that the travel and transport industries may either simply cater to the majority of travellers or may "prefer" single O/D trips. For surface travel, not being able to plan chained trips is out of line with the transport sector's policies to encourage multipurpose trips in order to reduce the number of trips and the kilometres travelled, to ease traffic congestion in cities and to lower the impact of transport on the environment. And in parallel, it should be possible to plan trips using multiple motorized modes (i.e. car, transit, or shared mobility) rather than primary mode only (i.e. car *or* transit).

Bias toward individual vehicle travel

Mobility data typically focus on what most companies care for, which is private vehicle travel – as per Google motto, "Waze is 100% powered by users, so the more you drive, the better it gets". Google Maps and like-structured apps have a built-in bias toward automobiles and trucks. Expressways are clearly marked at the regional and city levels. Street names appear at the neighbourhood level but not at the intersection level – which means that they are hard to use while walking in a dense city. This may be less a modal bias than a difficulty to display the rich and very complex environment of slow movement in a small handheld screen. Yet, and pointing to the possibility of a deliberate prioritization of individual vehicle travel, road tunnels are duly marked on the maps, even in highly walkable parts of cities (e.g. Les Halles in Paris and Pike Place Market in Seattle). Whatever the reasons behind the bias, it ends up supporting individual vehicular travel rather than active or public-transit-based transport, which runs counter to strategies to downsize the environmental impacts of transport and to support active travel.

Confusing social data with commercial data

Google points of interest have a bias toward commercial entities, which some will argue is a by-product of surveillance capitalism. Institutional or educational entities are less well represented and art facilities clearly underrepresented. The trend is likely to worsen as companies like Waze let businesses be listed on their app. One option would be for public-sector regulators to demand that companies such as Google or Waze freely list all public and cultural services on their maps.

Changes in neighbourhood travel

m-ICT has a direct effect on cities and on neighbourhoods, in particular, which is beginning to concern urban planners. On one hand, algorithms used by location-based services (e.g. Waze, Google) often prioritize local streets with low traffic volumes because their use leads the quickest journey. The phenomenon has not escaped local street abutters who are irate about noise and averse to what they consider speeding traffic. On the other hand, the rise of e-commerce has changed the nature of shopping trips. At this time, e-commerce comprises almost 18% and 15% of retail activity in the United Kingdom and the United States, respectively. Vacant storefronts are becoming common in many cities, threatening the quality (i.e. liveliness, safety) of future street environments for pedestrians. Furthermore, e-commerce generates new freight traffic on city streets, which raises many questions: To what extent does freight substitute for individual shopping trips? How is this trend affecting traffic in metropolitan regions? How can the efficiency of e-retail delivery be improved? Should the freight industry be better regulated and taxed differently? Should e-retailers be transparent about their transport costs? Should packaging and recycling costs be considered? These and many more questions must be answered in order to rein into the actual costs of transport related to the growth in e-commerce and to make a dent into the sustainability of current and future characteristics of shopping behaviours. The big data handled by mobility and location-based service providers could also serve to model how cities, neighbourhoods, streets and public space can be restructured and redesigned to accommodate the new behaviours efficiently and to enhance the quality of urban life.

Turning the tables

As smartphones can now yield quasi-perfect temporal and spatial information on people's travel, there are no reasons why public transport agencies cannot adopt data collection systems that mimic those of TNCs and location-based apps or why private companies could not benefit from crowd-sourced data by and for the "responsible traveller," the one who also uses and even perhaps primarily uses collective and active modes. As discussed earlier, data inclusive of all travel modes are needed for monitoring trends and modelling supply and demand. Models of comprehensive travel data collection already exist as, for example the sophisticated integrated transport payment systems used in Denmark. Some transport agencies collect their own data via smartcards (such as the Oyster and other smartcards used by Transport for London). Other agencies now more routinely administer their surveys through smartphones, collecting week-long data on travel. Furthermore, given advances in sampling and machine learning algorithms, it would be possible (and inexpensive) for public agencies to recruit small but representative samples of the population and have them share their Google location history data at regular intervals (Ruktanonchai, Ruktanonchai, Floyd, & Tatem, 2018).

Two-way communication with travellers in public transport

To remain an attractive alternative to TNCs services, public-sector transport systems need to improve ways in which they can interconnect with their users. As an example, the 2018 transport employees strike in France had minimal impact because the SNCF (Société nationale des chemins de fer français, France's national railway company) was able to keep travellers informed of schedules and schedule changes via Facebook and other social media platforms. Text service from transport agencies also seems to become increasingly available. By personalizing communication, smartphone-enabled information can transform how individual travellers relate to and experience collective public transport.

Conclusion

Smartphone technology is bringing us close to collecting mobility data that link individual travel demand with transport options in real time. The technology is currently driven by commercial interests which risk short-changing efficiencies in transport systems. Travel data must serve to optimize mode and route choice. Yet in today's m-ICT-enabled transport, transit and non-motorized travel options remain underrepresented, thereby compromising efforts to reduce the environmental impacts of transport and to promote healthy lifestyles. As McLuhan suggested, the need is for the technology to serve travel and not travel to serve the technology.

References

Campbell, S. W. (2013). Mobile media and communication: A new field, or just a new journal? *Mobile Media & Communication*, *1*(1), 8–13. https://doi.org/10.1177/2050157912459495

Carr, N. (2019, January). Thieves of experience: How Google and Facebook corrupted capitalism. *Los Angeles Review of Books*, January 15, 2019. A review of Zuboff, the Age of Surveillance. Retrieved from https://lareviewofbooks.org/article/thieves-of-experience-how-google-and-facebook-corrupted-capitalism/#!

The Economist. (2018, May). The Son kingdom: The impact of Masayoshi Son's $100bn tech fund will be profound: Briefing section. May 10, 2018. Retrieved from www.economist.com/briefing/2018/05/10/the-impact-of-masayoshi-sons-100bn-tech-fund-will-be-profound

McLuhan, M. (1964). *Understanding media*. Toronto: McGraw-Hill.

Ruktanonchai, N. W., Ruktanonchai, C. W., Floyd, J. R., & Tatem, A. J. (2018). Using Google location history data to quantify fine-scale human mobility. *International Journal of Health Geographics*, *17*, 28.

Zuboff, S. (2015, March). Big other: Surveillance Capitalism and the prospects of an information civilization. *Journal of Information Technology*, *30*(1), 75–89. Retrieved from https://cryptome.org/2015/07/big-other.pdf

Zuboff, S. (2019). *The age of surveillance capitalism: The fight for a human future at the new frontier of power*. New York, NY: Public Affairs.

Index

Note: Page numbers in italic indicate figures and page numbers in bold indicate tables.

Printed in the United States
by Baker & Taylor Publisher Services